abc
of the
Telephone

McCarty/Hooper

Volume 16

THE FINE ART OF

FAULT LOCATING

abc abc abc abc abc abc abc
abc abc abc abc abc abc
abc abc abc abc abc abc
abc abc abc abc abc abc
abc abc abc abc abc
abc abc abc abc abc

abcTeleTraining, Inc.

Box 537, Geneva, Ill. 60134 USA
Phone: (708) 879-9000

Books published by abc TeleTraining

Basic Series
abc of the Telephone, Vol. 1—Telephone theory, principles and practice
abc of the Telephone, Vol. 2—Station installation and maintenance
abc of the Telephone, Vol. 3—Central office plant
abc of the Telephone, Vol. 4—Outside plant; engineering and practice
abc of the Telephone, Vol. 5—Cable, inside and out
abc of the Telephone, Vol. 6—Understanding station carrier
abc of the Telephone, Vol. 7—Understanding transmission
abc of the Telephone, Vol. 8—Transmission systems
abc of the Telephone, Vol. 9—Design background for telephone switching
abc of the Telephone, Vol. 10—Principles of switching
abc of the Telephone, Vol. 11—Data communications practice
abc of the Telephone, Vol. 12—Practical grounding; theory and design
abc of the Telephone, Vol. 13—Grounding and bonding
abc of the Telephone, Vol. 14—Power line interference; problems and solutions
abc of the Telephone, Vol. 15—A basic guide to 1A2 key telephone installation
abc of the Telephone, Vol. 16—The fine art of fault locating

Traffic Series
Tables for traffic management and design—Trunking
Elementary queuing theory and telephone traffic
Principles of traffic and network design
Teletraffic concepts in business communications

Specialized Series
Anatomy of telecommunications
Telecabulary2 (An illustrated telecommunications vocabulary)
Voice communication in business
Basic telephone installation
Microwave facilities and regulations
Management in action
Noise investigation flow charts

PocketGuide Series
Noise reduction
Station protectors and how they operate
Grounding
Understanding PCM
C^3 Acronyms
Shield continuity testing
Telephonese Two (acronyms)
Telephone set circuits
Precise tone plan for end offices
Principles of party line station identification
Understanding static electricity
Auxiliary tip party marking devices

UNDERSTANDABLE TECHNOLOGY™

Published by abc TeleTraining, Inc.

Copyright © 1989 by abc TeleTraining, Inc.

Library of Congress Catalog Card Number 73-85629
International Standard Book Number 1-56016-041-1

Printed in the United States of America

9/89

98 97 96 95 94 93 92 91 90 89

12 11 10 9 8 7 6 5 4 3 2 1

THE AUTHORS

Donald E. McCarty is a specialist in the field of outside plant fault and cable location and in training technicians and managers for both national and international telcos. His effective straight forward approach to the subject is well established in the industry.

His utility career began in 1965 as an outside plant technician with Northwestern Bell. He worked in the line crew and as a facility technician doing splicing and cable repair, underground cable locating and air pressure maintenance. He also served as a testing technician, becoming familiar with all types of central offices and all types of cable plant.

In 1972 he joined the Dynatel Corp., traveling nationwide training both technicians and management in the use of Dynatel test equipment. From 1972 to 1979, he perfected several fault locating concepts, including "Section Analysis" and the "First Man on Site Concept." He also perfected a training course for underground conductor location (path and depth).

In 1977 he began formal management and craft training at Dynatel with four managers from Michigan Bell. The course has since become the industry standard. Procedures from the training course—from prelocalization of a single pair fault to section analysis—are now used by all operating telephone companies and have resulted in savings of millions of dollars in labor and disrupted service costs.

He was involved in the development of most of the existing Dynatel product line and is familiar with any problems the technician may encounter with test sets or any situation in the field.

When 3M purchased Dynatel in 1979, he continued management and craft training along with his other responsibilities as product sales manager and area sales manager for 3M until March 1986.

He left the company to begin training technicians and managers full time. In December 1986, he formed Donald E. McCarty and Associates Inc., an international utility consulting firm.

Glynn Hooper has been a free lance writer specializing in telecommunications since 1981. During that period he has produced articles, technical manuals and audio-visual presentations for vendors of telecommunications products.

He began working closely with Don McCarty as an outside contractor for the 3M Test & Management Division, supervising the writing and illustration of operating manuals and sound-on-slide audio-visual support shows for Dynatel test equipment.

The two have collaborated since on numerous projects, ranging from free-lance articles and a monthly column to test equipment evaluation and marketing studies for various vendors.

Hooper is president of Hooper & Associates, a technical writing and audio-visual production company. He has a BA and MA in English/writing from San Francisco State University.

Donald E. McCarty

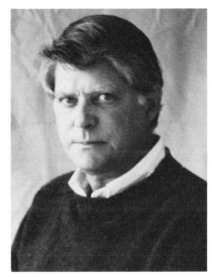

Glynn Hooper

Acknowledgements — The authors and the publisher gratefully acknowledges the contributions of several good telephone men which went into this book. There are too many to mention, but a special thanks to Bob Sanchez, Pacific Bell; Mike Townes, Dennis Slattery and Dennis Matthews of Southern Bell; Fred Zimmerman, Mountain Bell; and PIC cable; without whom this could never have been written.

This manual continues a long-standing tradition of *Understandable Technology*™ at *abc TeleTraining*. Built on the sound foundation of proven excellence in telephone training, the *abc of the Telephone*® manuals have been recognized as training classics for half a century.

The first volumes, originally written and published by Frank E. Lee beginning in 1942, soon became the standard for making technical concepts clear and easy to grasp. They established a basis for understandable technical training that is unsurpassed.

The abc series covering the spectrum of telecommunications remains the only complete series of technical educational texts and material available in the telecommunications industry today. In clear, conversational language, it stands as a basic introduction to telephone technology which, though undergoing constant change and development, retains its essential fundamental principles of operation.

Not only is this book a working tool and valuable reference on the subject, but it also can be used as an individual training guide. The abc series has become synonymous with understandable technical telephone education, whether self-study or formal.

Several new features contribute to the improved usefulness of the training manuals for self-directed learners:

1. Comprehensive outlines establish the parameters of coverage at the start of each chapter, to help the reader better organize and put into perspective the information to be studied.
2. Realistic, performance-based objectives set goals for the student and reader.
3. Preview questions arouse curiosity and give purpose to reading.
4. The list of technical terms provided for the reader to define and remember helps the student learn the language of telephony.
5. The review questions at the end of each chapter provide an opportunity for self-checking and reinforce the ideas once more.

In addition to its use as an independent training aid, but it can also be used in group training, for these same features enable instructors to prepare for and structure classes. The abc manuals have long been used in school and corporate educational programs throughout the world.

The increasingly universal acceptance and use of the abc TeleTraining publications and learning aids stands as a continuing memorial to Frank E. Lee, who wrote the first four manuals in the *abc of the Telephone*® series. Frank Lee dedicated his basic work *to the belief that progress in communications will help bring a better life to all people in every land throughout the world*. We remain dedicated to that objective at *abc TeleTraining!*

Joseph J. Aiken, Publisher

abc TeleTraining, Inc.

How to use this abc manual to help you learn!

Use this combined manual and student guide with the understanding that you are the instructor, the student, and the scorekeeper. Your success depends entirely on you and your determination to achieve.

Each chapter in this training manual consists of many parts: Outlines, objectives, preview questions, terms to be defined, review questions and answers, and brief comments.

We suggest you use this training manual and student guide as follows:

1. Read the outline of the chapter to get a sense of the organization of the content.

2. Read the objectives for the chapter. You will note that these objectives do not include such phrases as "... develop a better understanding of ...." or "... learn to appreciate ...." Rather, they state what you, the reader, will be able to *do* after completing the chapter. You may find that you need to go into greater depth. If so, you should plan on doing additional research and study. On the other hand, you may decide that being able to do some of those things listed is not particularly important to you. In this case read rapidly, skim or skip the chapter entirely.

3. Try the preview questions, and after you have completed the chapter, check back over your answers to see if you would answer them differently.

4. Read the chapter text. If the manual is yours to keep, we suggest you underline and make comments in the margin. If the manual is to be shared with others you don't have this luxury; take notes on a separate sheet of paper. Please understand that the goal is not simply to have read each paragraph in the manual—the goal is to know what each paragraph says.

5. Define the terms in your own words, then check the chapter to be sure of your accuracy and understanding.

6. Do the review questions. All of the questions and problems can be answered by you if you have a good grasp of the material in the particular chapter. There is also material in the text that is not tested in the review; so it is not sufficient to read just the problem, find the answer by referring to the text, and assume you have a total knowledge of the material in the chapter.

For some questions you will want to write the answers; for some this is unnecessary. Check your answers. You must judge for yourself whether you have enough correct answers, indicating a good understanding of the chapter. Or if you're part of a class, your instructor will tell you how many questions you should have correctly answered.

7. The comments that are included with the answers for each chapter at the back of the manual pertain to the particular questions for that chapter. In some cases they explain the answer; in some they simply expand upon the answers. In some cases no comments are necessary.

Good luck and happy learning.

THE FINE ART OF FAULT LOCATING

CHAPTER **1**

OUTLINE

What to look for in finding faults

Evolution of fault locating

The Wheatstone Bridge

Using audible tone

Dangerous innovations in service restoration

The breakdown set

Introduction of Plastic-Insulated Cable (PIC)

Fiber and the future

OBJECTIVES

After completing this chapter, the student will be able to:

1. Explain the test methods used in the early days and what their limitations were.

2. Describe what adjustments in testing were necessary as cable was improved.

Fault Location as an Art

PREVIEW QUESTIONS

As you read, watch for the answers to the following important questions:

1. What kinds of information and knowledge must a technician have to perform well at fault location?

2. What is the Wheatstone Bridge?

3. How many methods of fault locating exist? Is any one technique best?

Most telephone people understand that telephone conversation is made possible by sending a pulsating DC signal down a set of parallel wires. We submit that this is not entirely true. It is made possible by a form of witchcraft. All the rest is bureaucracy.

With this in mind, we have defined the process of fault locating as an art, rather than a science. Science involves working within a procedure governed by immutable laws which can be discovered, defined, understood and used with unerring confidence. Art adds intui-tion, instinct, talent and sly guesswork to the process.

The modern telephone technician works in an outside plant which has been growing and evolving for more than a century. Almost everything which was once in use is still in use somewhere. The technician must have a basic and thorough knowledge of the types of plant and the techniques needed to analyze and repair trouble.

What to look for
This technician must know:

1. What's out there—Including types of cable and the standards to which they are manufac-tured; types and techniques of construction and installation.

2. What goes wrong with it—Including types of faults; theories of resistance and capacitance.

3. How to find the trouble—Including section analysis; test sets (cords, dials, buttons); underground conductor location.

4. How to fix the trouble—Including techniques for long term restoration and recommendations for permanent repairs.

Since the invention of the telephone in 1883, someone has had to be around to repair it when it is broken. As the growing industry developed, techniques for repair also evolved. Open wire circuits required one type of expertise, underground and aerial lead cables another, and Plastic Insulated Cable (PIC) still another.

Evolution of fault locating

Perhaps the oldest method of locating a resistance or open fault on an exchange open wire circuit was simply to look for it physically. A repairman would follow the route looking for damage.

The Wheatstone bridge

To expedite repairs on toll circuits between towns, toll testmen used a "Wheatstone Bridge" (the grand ancestor to all modern resistance bridges) to isolate the fault. A distance to a resistance fault (called a "Varley," a measurement of resistive imbalance) such as a short, ground, cross or battery-cross, would be measured in ohms and then converted to an electrical-equivalent footage, based on the gauge, composition, and temperature of the wire. The toll testmen could then determine in whose area the fault lay and indicate to the appropriate outside repairman approximately where the fault was along his route.

A toll testman normally would measure the distance to fault in ohms and compare this figure to the known pre-tested electrical distance to telephone poles along the circuit. This would expedite the location of a fault and repairs could be made quickly.

This technique was demonstrated in our early years of cable repair, when working on an open wire toll route that served the two towns of Elk River and Anoka, Minnesota. After calling the toll testman at Anoka from Elk River, we strapped the faulted conductors as he instructed. After an agonizing period of time standing in hooks on the telephone pole, he indicated that the fault was 24.23 ohms from the Central Office (CO). (Because large diameter wire has relatively low resistance, measurements had to be exact to 100th of an ohm.) Allowing for wire diameter, temperature and composition, he converted ohms-to-feet and indicated the fault to be approximately 2.87 miles from the central office at pole #76.

Because of the testman's certitude, we asked what he thought the problem was. He told us that from the measurements and the magnitude of resistance on the meter, it looked like a hawk had become tangled in the wire. After removing the hawk at pole #76, we asked him how he could have been so sure. He said he had seen the hawk in the wires on the way to work that morning.

A technique called a "Murray Measurement" was used at the toll testman's test desk to measure the distance to an open by applying a tone to the circuit and adjusting the voltages on the wires until the tone balanced and cancelled. From this information, the capacitive difference in the wires was calculated to indicate distance to the open.

These "Open Wire" fault location techniques were carried over to lead-sheathed paper and pulp cable, and the problems encountered were many. With a Wheatstone Bridge, a clean, good pair was necessary to measure any resistance fault in these cables. Such a pair was rarely available because moisture would quickly saturate all pairs in the cable. This rendered the bridge almost useless.

Using audible tone

A method of locating such faults with an exploring coil was developed. An audible AC frequency was transmitted over the faulted wires, which acted as a kind of broadcast antennae.

The receiver was an exploring coil which was run along the cable until the tone stopped or noticeably diminished at the fault. If the fault proved to be a section between manholes (tone at one, no tone at the next), the bridge could once again be used with a clean pair strung along the ground.

For long and loaded cable, the coils were tuned to the 20-cycle AC current from the ringing generator in the CO. But this current would carry beyond the fault in short runs or in non-loaded cable, necessitating the use of test sets with different frequencies and voltages (from the original 91A to the 20C with the 105-type exploring coil). But because of tone carryover and power influence, faults were difficult to locate. When tape-armored cables were placed, tracing current was shielded, making all but solid faults impossible to find.

The solution was to make all resistance faults solid by literally welding (burning) the trouble pair together at the fault. In this situation, tone would stop at the solid fault and could be detected through any cable shielding.

Dangerous innovations

Service restoration required innovative, sometimes drastic and dangerous techniques. Wires were thrown over streetcar tracks and trolley wire to provide the more than 500 volts DC needed to burn a pair together. Tip and ring were welded into a solid short or cross, and an audible tone was applied across the conductors. A receiving coil and an amplifier were tuned to the same frequency. The technician would follow the tone until it stopped at the fault, and then make repairs.

Trucks were loaded with banks of batteries strapped in series that could provide as much as 1500 volts DC. The repairman would pass a set of jumper cables through the window of a central office, where a frameman would attach them to a telephone pair. The repairman would then take a set of carbons, briefly touch them together to complete the circuit and then draw an arc by separating the carbons. The larger the arc, the better the weld. Again, tone was applied and followed to the fault.

These dangerous burn techniques were refined over the years for several reasons. First and foremost was safety. An uncontrolled voltage applied to a telephone circuit could destroy telephone plant and customer equipment, endangering both customers and employees. Burning to ground was uncontrollable and outlawed from the beginning.

Innovative employees soon realized that if an ohmmeter were used to measure the short, the fault location could be calculated using formulas similar to the toll testman's work with the Wheatstone Bridge. The ohms could be converted to a footage based on wire diameter and temperature. They used manufacturer's specifications and the American Wire Gauge (AWG) charts for calculations.

> **EXAMPLE:** A technician measured a 100-ohm loop. He would divide the loop ohms by two for a single conductor's electrical measurement of 50 ohms. The 50 ohms times 38.54 feet-per-ohm in 24 gauge at 68 degrees would indicate the fault at 1927 feet.

The breakdown set

As the industry added more and more complicated central office and customer equipment, better control of voltage was required. A standard "Breakdown Set" was devised which used 14 45-volt batteries in series to produce 630 volts DC. A reverse switch allowed the repairman to reverse the polarity of the breakdown voltage. Reversing the direction of the voltage heated both ring and tip conductors at the fault, making it easier to burn the conductors together. A 577.5 Hz tone with an interrupter was built in to provide intermittent (for fault locating) or continuous (for coiling splits) AC tone. External battery posts were provided for adding battery to a Wheatstone Bridge to find light faults above 20 k ohms. A chart was installed in the lid giving the approximate feet per ohm in 19, 22, 24, and 26 gauge.

The breakdown set was the primary fault locating tool in the industry until the early 1950s, and is still used in pulp cable today. The invention of Plastic Insulated Cable (PIC) has rendered it useless in modern plant.

Enter PIC

When PIC cable was introduced, many people believed it to be the salvation of the industry. "Water doesn't hurt PIC," quoted our cable repair foreman. "When they manufacture it, they cool it in water." This turned out to be one of the many partial truths we dealt with in cable repair. Water doesn't hurt pulp cable either; not unless you intend to use it for service. Then it's a combination of water and central office battery that creates the problem.

Our PIC plant began to deteriorate, but at a very slow rate. Water would enter the core of the cable and some six months later we would see high-resistance, noisy-type trouble.

If water filled up a splice, the VOM (Volt Ohmeter) indications (grounded shorts and cross battery) were similar to a pulp wet. We would try to use the breakdown to burn the trouble. When that didn't work, we would cut the pair in half to see which way the fault was. Eventually, we would isolate the fault to a buried section of PIC. After several hours of group deliberation by the entire cable repair crew, someone would remember where an old trouble splice was. We would then look for indentations in the ground, old scrap wire, or anything that would indicate splicing activity. Once the splice was exposed, everyone held his breath until it was opened and water ran out. Then we would have a celebration. This was hardly efficient fault locating.

New filled cables are gradually replacing air-core buried cables, eliminating water problems in the underground. Early filled cable was not used in aerial plant as it tended to melt.

Recent innovations in filling compounds allow aerial use, but the cable is expensive and it is not yet widely installed. There are still millions of miles of problem air-core cable nationwide, both aerial and buried, which cannot be replaced just because there is trouble in a section.

The Delcon Corp., was the first in the field with a PIC bridge. The 4912F was a very sensitive resistance bridge that required both expertise and magic to operate. It was sensitive to both foreign DC and induced AC voltage. Static electricity would render it useless, unless it was placed on a splicer's box on a piece of slicker or fish cloth to insulate it from earth. Even then, a pencil eraser was needed to turn the dials without interference.

The principles introduced by this set were refined and advancements in technology have introduced a raft of accurate, easy-to-use resistance bridges and open meters, as well as microprocessor-driven test sets which examine all aspects of telephone transmission in the copper environment.

It appears that each time our testing and repair techniques approach a high state of efficiency in the field, the plant environment is changed, and we begin again. But no change takes place all at once. In Minneapolis, Minnesota, there is a layered, paper-insulated, lead and armor-sheathed cable under the Mississippi river which was placed in the late 1800s. This is still maintained. In rural areas throughout the nation, open wire circuits are still being installed and maintained. Well over half of metropolitan plant is paper and pulp insulated. This must be maintained. Plant everywhere has become a hodgepodge of cable types and construction as it is upgraded, repaired and replaced.

Fiber and the future

Now comes fiber optics—the wave of the future. It is obvious that test sets and techniques developed for fault location on copper conductors will not function in the fiber optic environment. An entire new set of theories, procedures, and testing equipment is evolving for this new fiber world.

Today's technician needs to be trained, not only for tomorrow's fiber optic world, but in each type of existing plant that is working in the field today. It's logical to buy the best plant in the world. If it isn't maintained, it's soon useless. It's logical to buy the best test equipment in the world. If it isn't used properly, it does no good. It's logical to develop the best techniques for an efficient, innovative telephone company, but if they are not used knowledgeably in the field, they won't work.

It is a matter of training and continuing education at all levels.

This manual is for the technician in the field and anyone else responsible for outside plant. The fine art of cable fault location in a mostly copper environment must be thoroughly understood.

As insulated wire is the basic unit of outside plant, a brief description of the manufacture of telephone cable is very important. If the technician understands both how a cable is manufactured and the built-in parameters required for quality service, subtle changes in these characteristics indicate the type of cable that has failed and proper repair procedure.

TERMS TO REMEMBER: *(Write the definitions in your own words.)*

Varley—...
....

Murray measurement—...
....

Breakdown set—...
....

REVIEW QUESTIONS:

1. What requirement in the use of the Wheatstone Bridge detracts from its usefulness?
A. The need for a clean, good pair to test against.
B. The need for mathematical computations by the testman.

2. With what kind of cable is the breakdown set still used today?
A. Pulp.
B. Lead-sheathed.
C. Plastic-insulated.

3. What set became useless in modern plant after PIC was introduced?
A. Wheatstone Bridge.
B. Breakdown.
C. All of the above.

(Answers on page 89)

CHAPTER 2

OUTLINE

Manufactured characteristics in cable

The manufacturing process

Types of insulation

Paper-insulated conductors
- Spiral-wrapped paper
- Longitudinal-wrapped
- Pulp insulated

Plastic-insulated conductors
- Low-density polyethylene
- High-density polyethylene
- Polypropylene
- Foam insulation
- DEPIC

Pair twists

Unit makeup
- Layer-type
- Unit-type

Special constructions

Functions of the cable machine
- Shield
- Sheath

OBJECTIVES

After completing this chapter, the student will be able to:

1. Analyze the characteristics of different types of cable.

2. Describe the manufacturing process and cable construction.

3. Discuss the reasons for development of different types of cable.

How Cable is Manufactured

PREVIEW QUESTIONS

As you read, watch for the answers to the following important questions:

1. What are the various steps in the cable manufacturing process?

2. What kinds of paper insulators have been used?

3. What different kinds of cable constructions are still in use?

In order to understand the principles of outside plant fault locating, it is essential that the technician understand how cable is manufactured and tested for acceptable customer service.

The technician must first understand transmission characteristics guaranteed by the manufacturer and tolerances accepted by the telephone company.

A technician depends on a precision test set to analyze a cable fault. Many of these test sets can measure resistance and capacitance to within 1/10th of 1% on a single unencumbered conductor. A resistance bridge with this accuracy could measure faults to within a foot per 1000 feet of cable. But because cable is built using various techniques to control transmission problems (such as the variable twist factor to eliminate crosstalk between cable pairs), test measurement accuracy is reduced by a factor of 10. This decreases any resistance measurement to an accuracy of one foot in 100 feet of cable.

Because capacitance characteristics of a cable pair can vary from a low of .076 micro-farrads/mile to a high .090 micro-farrads/mile and still be acceptable to the customer, a capacitance bridge is more difficult to use with confidence. By knowing of this variance, the technician can take steps (i.e., calibrate to a known length) to increase the accuracy of open measurements.

First, the technician must know the parameters acceptable on a nonfaulted working cable. By analyzing changes in these characteristics, an experienced technician will know the type of cable that has failed, the proper test equipment to use to analyze the type of fault, and where the fault occurs in the cable plant. After service is restored, the transmission quality must be tested and found acceptable.

A glance at the manufacturing process

Cable manufacture begins with 3/8 or 5/16 drawn rod shipped in from the smelting plant. It arrives by boxcar or truck where it immediately is redrawn down to a 10 to 13 gauge wire. As drawing work-hardens the wire, it must be annealed. This process heats up the wire and allows it to become soft and pliable again. The wire is then drawn to the proper gauge, annealed again, and prepared for insulating.

When covered with the proper dielectric (insulation), the wire is tested for insulation resistance, coaxial capacitance, and conductor resistance. It is then stored, ready to be made into a pair on the twining line.

On the twining line, individual conductors are twisted into pairs, tested and stored for the stranding operation.

In stranding, individual pairs are made up in sub-units or units, depending on the customer's requirements. The pairs are placed on the unit strander in the proper order and wound onto a take-up reel.

The take-up reel places the units in the proper predetermined order for good capacitance balance for transmission characteristics. The units are tested and sent to the cabling line.

The cabling machine puts a "lay" in the unit for flexibility and concentricity.

After testing the cable, a shield and a sheath are placed as determined by the customer for environmental protection.

Types of insulation

When open-wire circuits became impractical due to the immense growth of urban service, conductor pairs were grouped in close proximity in a sheath. A reliable insulation had to be developed to isolate each conductor from its neighbor. The requirements were many.

First, the insulation needed to have a high dielectric strength (the ability to withstand a voltage). To give a good coaxial capacitance for noise and power rejection, the insulation needed to be of a thickness equal to the total 360 degrees surrounding the copper conductor. The insulation had to stand up to the rigors of placing and be able to expand and contract under environmental temperature changes.

Paper-insulated conductors

Paper was the first insulator used as it was relatively inexpensive and in good supply. Three configurations of paper were used.

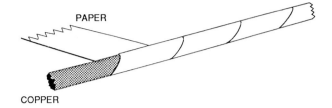

Fig. 1. Spiral-wrap paper insulation.

Spiral-wrapped paper. There were two forms of spiral-wrapped paper used. The paper was wrapped in a spiral around the conductor approximately 14 to 20 turns per foot (*Fig. 1*).

For capacitance control, the first technique was to wet the paper, then apply it to the conductor. As the paper dried, air pockets were formed. Since air has a lower dielectric constant than paper, conductor capacitance was lowered and loss was decreased. Telephone service could be extended without the use of a load coil.

The second technique was to double-wrap paper around the conductor. Two plies of paper lowered the capacitance by separating the pairs, and the added paper increased the dielectric strength of the insulation. This strength became a frustration for the technician, as anyone who ever tried to use a breakdown set on double-wrapped 22 or 19 gauge cable knows.

It seemed the only way to get enough voltage to break down the fault was to use two sets in series. This was a dangerous and illegal practice, but supervisors, knowing the difficulty, often ignored it. Such voltages could do itinerant damage to plant and facilities far removed from the fault.

NOTE: *When using a breakdown set, follow your telephone company practice to the letter. (In most operating companies it is located in section 634-305-501 of the practice or applications manual.)*

As spiral-wrapped conductors had a tendency to unravel during splicing, a longitudinal paper wrap was developed.

Longitudinal-wrapped. Longitudinal-wrapped paper insulation was not unlike using one-half inch paper tape to cover the wire (*Fig. 2*).

This method was acceptable early in the industry before power noise became widespread. Its big drawback was coaxial capacitance, because the thickness of the insulation varied. As larger cables were manufactured, noise became intolerable. New types of insulation with improved capacitance control were needed.

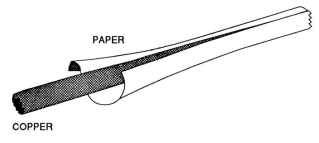

PAPER

COPPER

Fig. 2. Longitudinal-wrap paper insulation.

Pulp-insulated conductors. Paper insulation was used exclusively until about 1930, when pulp cable was introduced.

Pulp is made by pulverizing paper and beating it into a consistent slurry. This slurry, 99.95% water and .05% pulp, is colored in a holding tank. Three colors, red, blue and green, are used for the ring conductors (these are natural dyes and will not cause allergic reactions to cable splicers). The tip color is the tan of the unbleached paper.

The slurry is pumped from the holding tank to a cylinder mold where the water runs through, leaving the pulp on a fine mesh screen. The pulp is cut into strips and the conductor wire is pressed into it. The insulation is rounded out and dried down to a relative humidity of about 4%.

Plastic-Insulated Conductors (PIC)

Low-density polyethylene. Plastic insulation was introduced in the early 1950s. The first insulation was colored green-white and red-white, and toning for splicing remained the practice. As a carryover from paper customs, the extra pair concept of odd-count PIC was used for a period of time. This makeup had several disadvantages. Too many extra pairs wasted copper. As the same colors could occur on both tip and ring in the same count, splicers had problems remembering the color code and splicing errors were common. The technicians nicknamed this type of insulation "Wheel Cable," because a color wheel was needed to remember the pair.

Finally, an acceptable color code which always distinguished tip and ring was developed, and an even-count PIC was accepted. Because the colors available were limited to 10, cable was made up in 25 pair units (four units made up an even-count 100 pair cable). This material suffered from abrasion and pressure-slitting during manufacture causing leakage paths for moisture. Because of this, a high density polyethylene was developed.

High-density polyethylene. Newer plastics were used to eliminate the cracking, but other problems occurred over time. The next plastic used had an elongation problem and there was a tendency for the insulation to shrink back when the wire was cut for splicing.

Polypropylene. The last of the solid plastics used was polyproplyene. It was tougher than high-density polyethylene and could withstand heat better.

Foam insulation. With the introduction of filling materials to cables, the thickness of the insulation had to be increased to compensate for the increase in mutual capacitance. This was objectionable to the customer, because it meant less pairs per cable sheath and an increase in cost. As a result, expanded insulation was introduced. By adding gas to the insulation, the dielectric constant was reduced and the same insulation diameter could be used as the solid air-core cables. The insulation proved fragile and tended to pull off the wire when twisted for splicing. Because of this, DEPIC insulation was developed.

Dual Extruded Plastic Insulation (DEPIC). Expanded insulation is extruded over the conductor and a solid skin of high density polyethylene is extruded over the foam. The diameter is the same as solid air-core insulation. The cost is higher, but the handling characteristics are better.

Pair twists

The "twisting or twining" machine introduces a twist into the cable pair. Twists are staggered between one and one-half and six inches in a unit. In the older pulp cables there were nine different twist lengths, 10 if there was a "tracer" (blue-red) identification pair for tagging or testing the unit (*Table 1*).

Table 1. TWIST LENGTHS OF PULP CABLE (IN INCHES).

	19 GAUGE	22 GAUGE	24 GAUGE
Green natural black	4.5	3.9	2.6
Green natural orange	3.6	3.2	3.3
Green natural	2.7	2.7	2.0
Red natural black	4.8	4.1	2.8
Red Natural orange	3.9	3.5	3.4
Red natural	3.0	2.9	2.2
Blue natural black	5.1	4.3	3.0
Blue natural orange	4.2	3.7	3.6
Blue natural	3.3	3.1	2.4
Blue red (tracer)	5.4	5.1	3.8

In plastic insulation, there are 25 different twists in each 25 pair sub-unit (*Table 2*).

The pair twist provides balance between

Table 2. PIC AND FILLED CABLE. ALL UNITS INCHES PER TWIST.
MOST MANUFACTURERS

PAIR #	COLOR	19	GAUGE 22	24 AND 26
1	White blue	2.0	2.0	2.0
2	White orange	5.3	4.9	3.2
3	White green	3.2	3.1	2.7
4	White brown	5.9	5.2	4.1
5	White slate	4.1	3.9	3.7
6	Red blue	2.9	2.8	2.4
7	Red orange	4.5	4.1	4.5
8	Red green	3.8	3.7	3.5
9	Red brown	2.3	2.2	2.2
10	Red slate	5.5	4.9	4.7
11	Black blue	2.8	2.7	3.9
12	Black orange	4.9	4.5	3.1
13	Black green	3.3	3.2	2.6
14	Black brown	2.6	2.5	3.3
15	Black slate	5.1	4.6	4.2
16	Yellow blue	6.1	3.5	2.9
17	Yellow orange	2.1	2.1	2.1
18	Yellow green	5.7	5.1	3.8
19	Yellow brown	4.8	4.3	2.8
20	Yellow slate	3.0	2.9	4.0
21	Violet blue	3.9	3.8	2.5
22	Violet orange	4.7	4.2	3.0
23	Violet green	2.4	2.4	2.3
24	Violet brown	3.5	3.6	3.4
25	Violet slate	4.3	4.0	4.4

conductors and shield, and assures that the pair will always be equidistant from the shield. The varying twist length of adjacent pairs minimizes crosstalk and improves transmission quality.

Unit makeup

Unit makeup is important to the technician as it affects the accuracy of an open meter. Open meters designed for conformance testing are calibrated for nonworking cables. Those designed for fault locating are calibrated for working cables. When measuring opens in working cables, 12 or more pairs in the tested unit must be working or the indicated footage will read short of actual distance to the open. Unit identity is determined by a binder tie.

Layer-type construction. The first layer-type construction was concentric layer construction (three pair surrounded by 10 pairs, etc.). Pair twist in each layer was in a reverse direction from the preceding layer. As cables became larger, friction stiffened them. To offset this, several adjacent layers were twisted in the same direction.

Layer-type construction was used until about 1930 when unit-type construction was introduced.

Unit-type construction. Tested cable pairs are formed into 50 and 100 pair units. The unit is bound by two colored cotton binders. One binder indicates the gauge of the conductors (*Table 3*), the other indicates unit placement in the cable.

Table 3. GAUGE INDICATED BY COLORED BINDER IN PULP CABLE.

COLOR	BLU	WHT	RED	ORN
Gauge (AWG)	19	22	24	26

The strander puts a lay in the unit. This lay assures that the average pair length throughout a cable is consistent. All cable pairs are manufactured with a natural left-hand twist. Center units, which travel less than an outer unit, have a right-hand lay, while outer units have a left-hand lay. This makeup separates the pairs in adjacent units from unit-to-unit capacitance (crosstalk). The length of the lay is controlled by the take-up reel.

Pulp unit stranders were modified for PIC. Oscillating face plates were used for 25 pair sub-units. These face plates rotated back and forth from 180 to 360 degrees in a distance of 100 to 150 feet. This kept pairs in the sub-units separated from a standpoint of crosstalk. The sub-units could then be formed into 100 pair units on the same machine as was used for pulp cable manufacture.

Special constructions

Special designs were used, such as quadded cable for phantom circuits, broadcast circuits and carrier circuits. Video pairs were high-grade plastic with two layers of copper shielding. Coaxial units were designed to provide 1800 circuits per tube. This type of construction was also used for CATV.

Composite cables were used for dual purposes, such as toll (19 gauge "N" carrier)and broadcast (16 gauge "K" carrier) under the same sheath.

Cable machine

The rotating take-up reel of the cabling machine puts a lay in the entire assembly of units, so the cable may flex during placing and under temperature changes in the field.

When pulp units are made into a cable, the cable core is dried to a relative humidity of less than 4% immediately before the sheath is applied. The assembled core is spiral-wrapped with heavy paper to insulate the pairs from the shield and to protect the pairs from voltage such as power or lightning. The cabling machine places one complete lay every 42 inches.

When PIC units are made into a cable, the core is wrapped with mylar for protection. The average lay in a PIC cable is 36 inches.

Shield. The shield is added to the dried cable core. This is an essential electrical element for transmission characteristics. It stops the intrusion of induced AC and guides it to ground, it balances the capacitance in the presence of such AC, and gives the cable strength and protection.

Sheath. The cable sheath is the first line of protection for the cable core. Lead sheathing was an excellent protector, but costly and difficult to work with. Plastic sheathing was first thought to be the answer to moisture intrusion, but experience showed that the glue between the sheath and the shield was the actual moisture barrier. The plastic itself allowed moisture osmosis eventually. Because of this, several different sheaths were developed.

Underground cables needed one type of protection, while direct buried and aerial cable needed another. The sheath itself (unlike the shield and conductor insulations) is not a major factor in fault locating.

Outside telco plant is subject to all the forces of man and nature, which often seem to conspire against phone service. There are typical patterns and problems in cable failures which can lead to early analysis and permanent fixes. Knowing the cable and the troubles it is heir to will speed restoration.

We turn to a discussion of cable failures.

TERMS TO REMEMBER: (Write the definitions in your own words.)

Unit— ....

....

Dielectric strength— ....

....

Shield— ....

....

Sheath— ....

....

REVIEW QUESTIONS:

1. An accurate resistance bridge can measure faults per 100 feet of cable to within:
A. Six inches.
B. One foot.
C. One yard.

2. In plastic insulation, pair twist lengths per sub-unit are:
A. 25.
B. 35.
C. 50.

3. In order to measure opens in working cable, the minimum number of pairs working must be:
A. Eight.
B. 10.
C. 12.

4. A cabling machine places one lay every:
A. 25 inches.
B. 42 inches.
C. 60 inches.

5. The capacitance characteristics of a cable pair can vary from:
A. .076 micro-farads/mile to .090 micro-farads/mile.
B. .030 micro-farads/mile to .064 micro-farads/mile.
C. .045 micro-farads/mile to .087 micro-farads/mile.

6. The varying twist length of adjacent pairs minimizes:
A. Crosstalk.
B. Elongation.
C. Dielectric strength.
D. All of the above.

(Answers on page 89)

CHAPTER 3

OUTLINE

Pulp and paper conductor faults

Water in air-core PIC

Water in filled cable

Splice and encapsulation failures

Sheath damage

Common field problems in PIC
- Low-density polyethylene
- High-density polyethylene

OBJECTIVES

After completing this chapter, the student will be able to:

1. Describe common cable faults and what to expect when a cable fails.

2. Sort out the various causes for trouble reports.

3. Select the proper approach to correct the fault.

4. Recognize repair approaches which can lead to further customer complaints.

Cable Failures

PREVIEW QUESTIONS

As you read, watch for the answers to the following important questions:

1. What are the major causes of failure in various types of cable?

2. What is the result of water entering a cable and how can it be corrected?

3. What are the various types of PIC cable and what are the advantages and disadvantages in their use?

When a cable goes down, there is a reason for it. That reason may be—and far too often is—found by trial and error. Or the trouble may be solved—again, far too often—by cutting to clear and removing revenue producing copper from the system. An educated, systematic analysis of a trouble, though it takes a few minutes of thought, will put a customer back in service and find and fix the source of the problem. To do this, we must know the types and causes of cable failure.

Although cable plant is as tough and tight as we can make it, there are an amazing variety of troubles which occur. These range from acts of God (lightning, fire), acts of man (shovel damage, improper closures) to Mother Nature (rodent damage, water intrusion, etc.). In nearly all outside plant faults, there is, in the end, a failure of the protection required to assure proper current flow. When that protection fails, phones don't work right. Insulation surrounding a conductor has *the* major effect on trouble symptoms.

Following are examples of common faults in various cable types, and what to expect when a cable fails. We include practices now in use, which often invite repeat dispatches, and examples of how section analysis techniques can help eliminate this problem.

Pulp and paper conductor faults

Pulp- and paper-insulated conductors have a low dielectric strength and are susceptible to power and lightning. If a power line were to fall on a telephone cable, open conductors will occur where lowest insulation resistance allows a return path to ground, melting the conductors. The average bolt of lightning (containing up to 65,000 amps) does the most damage at its exit site—which may be miles from its strike. Protection of plant and customers requires good bonding and grounding procedures to divert lightning safely to ground. When such protection fails, the conductors may all melt together or even be completely vaporized by the heat. The entire cable may be affected in areas other than the initial damage, causing individual conductors to be shorted, grounded or open at random sites.

Service interruptions in pulp cable are most often caused by water intrusion. When water enters the cable, the cause and the effect are instantaneous, and the customer remains out of service until the water is removed.

After the pair is isolated from the CO and tested, the out-of-service pairs will test 48-52 volts DC from either tip- or ring-to-ground. Testing across the pair with no voltage present, a short from 2000 ohms to about 50,000 ohms will be indicated.

Such a pattern of symptoms indicates that most likely a pulp cable has failed. If automatic testing (such as MLT Predictor or GTE Fortell) does not show the magnitude of the fault, random testing of one pair per 25 will indicate the number of cable pairs involved and help localize the problem (for example, if 400 pairs show as wet, the problem is not in a 50 pair cable).

If a paper or pulp wet is suspected, the primary test set is the breakdown set. A bridge-type device is ineffective because normally a "good" pair cannot be found in the cable.

> **CAUTION:** When the breakdown set is used in any electronic office **ALL** associated **WET** cable pairs **MUST** be disconnected (by pulling the coils or disconnecting all working pairs at the cross-box) before breakdown voltage is applied. This will protect sensitive CO equipment which can be damaged by minute amounts of foreign voltage.

The breakdown set is three test sets in one. First, it supplies high voltage for "breaking down" (welding) a high resistance fault (above 1.5 k ohms) to a low resistance fault (below 1.5 k ohms). The breakdown set also supplies tracing current (577.5 Hz) to locate the fault with an exploring coil. The set has a meter to identify pair faults (when CO battery is disconnected) and to measure the resistance of solid faults. The ohms to the fault can be converted to an equivalent electrical footage based on the gauge and the temperature of the wire.

> **RULE:** The breakdown test set is not designed to be used on plastic-insulated conductors. Plastic insulation is designed to withstand a minimum of 2200 volts, which is more than the breakdown set can **LEGALLY** provide.

NOTE: The test set is not very effective on aluminum conductors. Its voltage melts the wire causing opens.

Water in air-core PIC

Plastic Insulated Cable (PIC) has a high dielectric strength and moisture intrusion can take place without immediately affecting service. It was once thought that PIC technology was the answer to moisture problems and that the extruded plastic covering would protect the conductor indefinitely even with moisture in the cable, as long as no outside force caused damage. But experience showed that faults will appear in a wet section after several months of immersion.

Most plastic insulation has microscopic air bubbles which are undetected by the spark tester during manufacture. After water has saturated the cable for a long period of time, the water will permeate the plastic at these thin spots and allow current leakage between the conductor and shield (*Fig. 1*).

A second and very subtle problem is induced AC on conductors. AC current flows around any solid impurities (minute grains of foreign material) in the plastic conductor insulation. This action stresses the insulation and eventually causes microscopic cracking, creating a pathway for moisture to reach the conductor. It may take as long as six or seven years for this to occur, depending on the severity of the induced AC. Once the cracks occur and water reaches them, there will be current

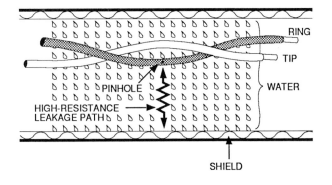

Fig. 1. Water in PIC cable.

leakage between conductors and shield.

Whatever the reason for the intrusion, water in PIC shows a pattern of symptoms which, when analyzed, will indicate water in a splice or encapsulation, water at the site of sheath damage, or water internally in air-core cable. Any combination of water and DC battery will result in electrolysis (the electrochemical decomposition of metal). If the water is internal, the shield will be affected. If the water is in a splice, encapsulation or damage spot, the conductors will be affected (Fig. 2).

A typical telephone system uses a negative potential of 48-52 volts on the ring, with the tip conductor at earth-ground. In reference to the ring, this makes both the tip and the earth-ground positive (Fig. 3).

The tip conductor and earth-ground are of the same potential, so no current flows and no electrolysis occurs between them when water enters the cable. The aluminum shield and the tip conductor are sacrificial (they lose molecules of metal) to the negative ring potential. As dielectric failures on the tip may or may not be in the vicinity of the faulted ring, electrolysis is usually confined to decomposition of the aluminum shield (Fig. 4).

Thus, testing a cable conductor-by-conductor with a VOM would show crossed battery on ring conductors with very few tip faults. The indicated battery would be very low voltage. Testing the isolated shield would show from 15 to 30 volts DC on the shield if there is no sheath damage.

The above symptoms confirm that there is water in the cable and no sheath damage. Resistance bridge measurements indicate faults at random footages throughout the water. Capacitance measurements on a known good pair show the section as longer than the actual length, indicating the amount of water. The Time Domain Reflectometer (TDR) analysis indicates the location of the water in the section.

SOLUTION: *Purge or replace the cable.*

An all-too-common *temporary* solution to water in PIC cable is to "cut to clear." The problem is in the testing: when battery and ground are removed from the pair, electrolysis ceases. Galvanic corrosion begins. The two dissimilar metals, aluminum and copper with an electrolyte (water) between them, create a battery effect. As aluminum is sacrificial to copper, the aluminum oxide, which is nonconductive, will

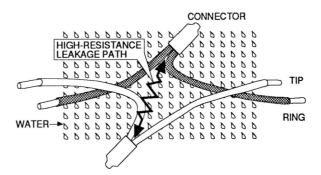

Fig. 2. Water in a splice.

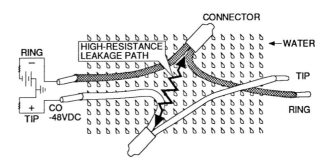

Fig. 3. Tip is sacrificial to negative ring potential.

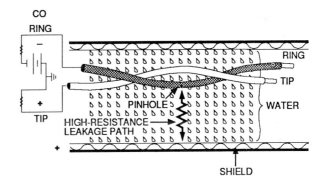

Fig. 4. Shield is sacrificial to negative ring potential.

coat the pinholes where the copper is exposed. Electrolysis temporarily ceases and the pair appears to test OK.

A technician uses a VOM to test a vacant pair to see if it's good and gets to the customer. The capacitance charge induced by the meter's internal battery indicates the length of a vacant conductor. This is shown by a momentary deflection of the needle which returns at once to zero. The longer the wire, the more the capacitance, and the more of a kick on the meter. Technicians interpret that kick in points, with each point indicating a footage.

When an inexperienced technician retests a pair which has been cut away from on an earlier dispatch, galvanic corrosion may have sealed the pinhole faults. The only indication of a problem is an extended capacitance kick on the meter. The meter will kick to mid-scale and temporarily remain there, indicating a resistance fault. But the meter starts to bleed the trouble off. Generated battery on the shield is bled through the meter to ground. All too often this reaction is encouraged by the technician: it's 3:45 P.M. and quitting time is 4:30. As the meter drops to zero, the pair appears clean.

When battery and ground is re-applied, electrolysis begins again, and static begins again, and the customer complains again.

> **RULE:** Do not use a tested vacant pair that indicates an extended kick on the VOM and then slowly bleeds off.

Cutting to clear is the quick fix which lowers dispatch time, and the trouble conductor will test good on a later dispatch. When it is returned to service on a future cut to clear, it will soon show the old trouble. This is not considered a repeat report only because the trouble pair is working on a different number. Same trouble, new number.

This scenario will be repeated over and over because inexperienced technicians are encouraged by constraints of time to cut over. Eventually, after spending a considerable amount of money on what amounts to a single repair problem (water intrusion), there is nothing left to cut to. The cable must be replaced.

Water in filled cable

Recent studies have shown a problem with the oxidative stability of PIC cable insulation. Depending on climate (the Southwest has the major problem), the Oxygen Inductive Times (OIT) of the insulated conductors exposed to air (in pedestals) have led to cracking of the foam-skinned, high-density polyethylene-insulated conductors sheathed in cable filled with waterproofing compound. The cracks occur over a period of time naturally, and with any handling of the brittle insulation they will increase. They admit condensation moisture to the conductors. This supports the case for enclosed or air-tight plant. All pair access in the distribution plant should be through fixed-count designated terminals. This will drastically slow down OIT and eliminate any single pair faults caused by technicians handling exposed cable pairs in terminals.

We have spoken with many technicians in the field, but we have *not* witnessed or heard of a cable fault directly attributed to water in filled cable sheath. Failures occur at splices, encapsulations and damaged sheath, and water may possibly travel a few feet from the intrusion point along disturbed conductors. Replacing damaged air-core cable with filled cable is a proven long-term solution to water intrusion in a cable section.

Splice and encapsulation failures

Water intrusion into a splice or an encapsulation may happen suddenly, or years could pass before troubles begin and the cause is recognized.

When this occurs, tested faults indicate grounded shorts with crossed battery on both ring and tip. Resistance faults between ring and tip are usually in the 100 k ohm range. Tip conductors will be "going open" due to electrolysis (*Fig. 5*).

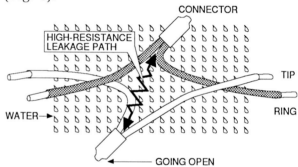

Fig. 5. Water in a splice case or defective encapsulation causes a high-resistance open on the tip.

There usually are plenty of pairs to cut to if the problem is in the underground. If the problem is beyond the work-out terminal, the pair is Cut Off Beyond (COB). A proven section fault is "cut around" until there is nothing left to cut to, then PBX or similar wire is laid on top of the ground to bypass the bad section. Trouble reports by subscribers indicate: "Cable cut by lawn mower," "Chewed by Pit Bull," and so on.

Each case of trouble must be taken on its own merit. Don't assume; prove it. There is a tendency to attribute a case of trouble to a "known bad area" and to cut to clear without doing much analysis.

A good example of this occurred in Kansas. When we were doing training on new test sets, we requested a known bad area for testing after giving classes to several repair technicians on the section analysis procedure. It was our intention to demonstrate the procedure in an actual field condition.

An isolated subdivision was fed from a control point to a cross-box by a 600 pair PIC cable. The trouble was already isolated to the plant beyond the control point. When we entered the subdivision, customers threw rocks at the telephone trucks, referred to our parentage with derogatory remarks, shouting "don't you touch my line!" Dogs snarled and mothers hid their children. From this, we concluded there was a problem that had existed for a while.

We isolated the fault between the control point and the cross-box, and determined that there were approximately 3500 feet of 600 pair, single sheath cable in the test section.

After shorting a pair between the two points, we set the gauge and temperature of the cable and measured the distance with a resistance bridge. The bridge indicated 3650 feet electrically (the extra 150 feet was due to the resistance of a loading coil). The open meter indicated 3500 feet of cable, which showed no water was present in the section.

From this information, we knew we were looking for either a bad splice or a damaged spot on the cable.

VOM testing on several faulted conductors indicated grounded shorts, crossed-battery, and the bridge strap-continuity test showed tips going open. Guess: bad splice.

Bridge measurements on several pairs indicated a fault 590 feet from the control point. This was confirmed by the earth-gradient frame.

Exposing the indicated trouble spot revealed a load splice with two 300 pair load coils. The splice was encased in a lead sleeve filled with "F" plugging compound. A pinhole in the sleeve allowed water to enter and migrate through the splice.

This analysis took approximately two hours from start to finish, including coffee.

A case history study indicated the first faulted pair was cut to a new pair in April 1968. Faults occurred at a rate of one or two a month from the time of the initial report until we exposed the load splice. This problem caused customer aggravation from 1968 to 1976, and a documented $1,000,000 + was spent only on the *effect,* not the *cause.*

A trained technician, who is encouraged to analyze and cure troubles the first time out, would have saved the company considerable money.

Sheath damage

When testing a section, sheath damage will have the same indications as a bad splice, except that solid or low-resistance shorts and opens will be indicated by the VOM. The process for identifying the trouble site is the same as when locating a bad splice or encapsulation.

Damaged cables can go unnoticed for an extended period of time without proper analysis and reporting. We saw an example of this in Florida.

Service technicians, installing service in a new subdivision, cut 105 pairs from their original 900 pair assignment before a cable repair supervisor was made aware of the problem. Investigation of the section revealed that over 200 pairs had been damaged during placing. By cutting to clear, over $20,000 was spent on the effect of this damage.

NOTE: *All direct-buried cables must be tested for sheath damage after placing and before splicing.*

Each customer report, properly analyzed and throughly pursued can eliminate unnecessary cutting. Through proper analysis, the first man on site can determine if the faulted pair is an individual one-pair fault affecting just that customer or if the fault is part of an eventual cable failure.

> **REMEMBER:** Water in the cable, water in a splice or encapsulation or cable damage can occur individually or together.

Common field problems in PIC cable

Today's field technician deals with a variety of PIC cables, from the earliest types to the present filled technology. Differing construction methods, untested materials from bad connectors to unacceptable terminals, and poor or uninformed installation techniques, exacerbated problems in this plant.

Low-density polyethylene. The first plastic was sensitive to heat, and field problems occurred. The accepted method of construction was to loop the cable into a pedestal, expose the pairs, and splice the conductors directly to the drop. An unvented "dome" type terminal was then placed over the pairs. In direct sunlight, the high internal temperature in this installation cracked the plastic insulation. There was no moisture protection placed, and when the terminal cooled in the evening, condensed moisture would enter the splice. Gradually, electrolysis caused the tips to go open.

The customer complaint was static or noise on the line. The pair would test good from the CO and telephones could be identified, but when the customer was called, the static could be heard. The only way to find this type of "battery open" was to listen first at the workout terminal. If noise was heard, the technician worked toward the central office. When the line was quiet, the technician had passed the trouble. Thus, by a slow process of elimination, the fault was proved to the affected terminal.

Even though this method of construction was abandoned, this type of terminal still exists in the field nationwide and is a constant source of trouble in ready access plant. In the terminal, the cable pairs are held by a thin sheet of mylar taped in place. Experience teaches the technician to very carefully unwrap the mylar without disturbing the splice; the plant here is so fragile that many other troubles could be caused by disturbing other pairs. Binder ties, which compressed the unit and caused troubles when the PIC cracked, were often removed and discarded on earlier dispatches, and bonding rings were lost. With no binder ties to indicate groups, pair identification is normally done by tone. The technician very carefully digs out his pair without disturbing any other pair. A tug on the tip side opens the pair. After repair, the technician will carefully rewrap the splice so as not to cause other problems. Time constraints forced the technician to clear the trouble pair, but leave the problem.

Experience taught the technician to inform the customer that service had been restored, ask for a glass of iced tea and retire to a lawn chair while the maintenance administrator tested all other pairs in the complement for troubles caused by this visit. These would be fixed. Future complaints in the same count could quickly be isolated to the same terminal.

High-density polyethylene. High-density polyethylene was introduced about the same time that aerial lead legs of cable had reached the end of their life. The lead cables were replaced with the newer plastic. The terminals were called "ready access" as all pairs were exposed to any technician who opened it. A designated pair was spliced to a block and the drop attached to the lugs. At the end of the cable run, about eight inches of sheath were removed and the pairs were Christmas-treed (cut at irregular intervals). An oversized cable cap was filled with "F" plugging compound and the cable was strapped down the pole.

With ready access (often called "rat's nest") plant, the majority of faults were man-made. A technician would cause service interruption while installing or repairing another circuit. Nationwide, for better than 20 years, these "single pair" faults caused some 75% of all customer complaints that were attributed to cable.

Most faults were cleared by isolation. The technician would cut the pair in half to see in which direction the fault was. This leapfrog technique could require seven attempts (worst case). If the fault was attributed to the cable, more than likely the end cap would be wet.

Most technicians worked on the *effect* of the problem, rather than the cause. The technician would open the pair at the Access Point (AP) where the cable pair left the underground run and entered the distribution side of the plant. If the fault was beyond the AP, the repairman would proceed to the work-out terminal and disconnect the drop to prove the fault in the drop or the cable. If the fault was in the cable, the technician would COB the pair (Cut Off Beyond the work-out terminal). A tremendous amount of time was wasted and customers were put to unnecessary aggravation.

Eventually a cable repair technician would be dispatched and discover more than one pair in trouble. The technician would see grounded shorts in the 100 k ohm range with crossed battery in the 15 to 30 volt range, but no open cable pairs. The technician would finally determine which multiple lateral was faulted by opening a pair at the beginning of the lateral.

The next step was to cut off the suspected end-cap. If the fault was still there, the technician would visually inspect the entire lateral, looking for a 4A-type closure that had rolled over and filled with water or a ready access terminal with a bird's nest in it.

Improvements were made both in plastics and construction techniques. Because of public demand, most new construction was buried even though the expense was greater. Water entered the cable through failed splices and damage sites caused during construction or by foreign workmen. Again the majority of faults were cut away until there was nothing left to cut to. A wholesale replacement of sections was used to restore service.

The general movement to buried plant, coupled with the ready access maintenance concept, proved air-core PIC surprisingly vulnerable to water intrusion and the slow, steady failure of pairs due to electrolysis. Filled cable was developed to stop water intrusion and has proven effective.

Ready access plant must be locked up and pairs accessed only through binding posts. The oxidative stability of PIC insulation is decreased when exposed to air and cracked insulations

will occur. The second benefit of locking up the plant is keeping hands away from churning the plant and causing one-pair faults.

Knowing what is out there; having a good idea of the physical characteristics, electronic capabilities, and system weaknesses will arm the technician with the education and tools to do the primary job: maintain the customer in service. Once a trouble is found and repaired, DC charactristics of the circuit are proven good. On the same dispatch, AC characteristics should be tested for transmission quality and unacceptable parameters corrected.

TERMS TO REMEMBER: *(Write the definitions in your own words.)*

Electrolysis—..

....

Cut off beyond—...

....

Access point—..

....

Ready access terminals—...

....

REVIEW QUESTIONS:

1. Where does the average bolt of lightning cause the most damage?
A. At its entrance site.
B. At midpoint.
C. At its exit site.

2. Service interruption in pulp cable is most often caused by:
A. Water.
B. Digging.
C. Lightning.

3. What is the primary test set if water in pulp cable is suspected?
A. Wheatstone bridge.
B. Breakdown set.

4. With the tip conductor at earth ground, a typical telephone system uses a negative potential of:
A. 36-42 volts.
B. 48-52 volts.
C. 59-68 volts.

5. What is an all too common solution for water in PIC?
A. Cut to clear.
B. Purge cable.
C. Replace cable.

6. In PIC resistance faults between ring and tip are in the range of:
A. 60 k.
B. 100 k.
C. 110 k.

7. Testing wet pulp across the pair with no voltage present, a short will be indicated from:
A. 20,000 ohms to about 500,000 ohms.
B. 2000 ohms to about 50,000 ohms.
C. 200 ohms to about 5000 ohms.

8. Testing wet air-core PIC will indicate crossed battery on:
A. The ring conductors.
B. The shield.
C. Both.

(Answers on page 89)

CHAPTER 4

OUTLINE

How to use a Volt Ohmmeter (VOM)

Factors affecting ohms measurement
- Gauge
- Temperature
- Wire composition
- Helical cable design

Resistance bridge operation
- Identify the faulted conductor
- Test the good pair
- Attach the far-end strap
- Gauge and temperature setting
- Measure distance from test set to strap
- Gauge change
- Slack loops and butt splices
- Loading coils
- Nulling: clearing and balancing the bridge
- Adjusting measurement for resistance factors

OBJECTIVES

After completing this chapter, the student will be able to:

1. Operate a Volt Ohmmeter.

2. Use a resistance bridge.

Resistance and Using a Resistance Bridge

PREVIEW QUESTIONS

As you read, watch for the answers to the following important questions:

1. What are the functions of a Volt Ohmmeter (VOM)? A resistance bridge?

2. For what kinds of troubleshooting are each of the above best suited?

Resistance may be defined as that property of a conductor which determines the amount of current flowing through it at a given voltage. The standard unit of resistance is the ohm.

When a service interruption proves into the cable, normally the fault is resistive (a short, ground, cross, or battery cross). When testing the insulation resistance, the technician can determine the type of cable faulted and the proper procedure for finding the problem.

A central office computer, such as MLT II or Foretell, tests, identifies, and pre-localizes a resistance fault. The computer has a bridge function that measures the linear wire resistance in ohms from a strapped terminal back to the fault. If the pair is vacant or there is no access to a CO bridge, a field bridge is used with a strap either in the CO or in an associated terminal.

How to use a Volt Ohmmeter (VOM)

Most operating companies are using an industry-standardized VOM. The meter has an AC scale (which uses a separate AC probe with a DC blocking capacitor), a DC scale, a resistance scale, and an mA scale.

The testing sequence is:

Connect the black test lead to the COM jack and the red lead to the V-ohms-m jack. Select the highest AC scale and test for hazardous AC or DC. This can be done with the DC probe. If voltage is indicated and needs to be measured, remove the DC probe and connect an AC probe. The probe's internal capacitor will block DC and measure AC. Use the lowest appropriate scale. If a dedicated AC probe is not available, connect a 1 μF capacitor in series with the DC probe.

If no AC is present on the circuit under test, use the DC probe for all further testing.

To test for DC, connect the black lead to

the test conductor and the red lead to ground. If the meter deflects left, the voltage is positive DC. Reverse the leads.

To measure current flow, connect the meter in series with the circuit. Use the highest range and down range to the lowest range.

> **CAUTION:** Always use the highest scale and down range for an accurate reading. Avoid pegging the meter as damage may result.

If no voltage is present, select the highest ohms range (normally R x 1 K). Move the selector switch to the range that gives a one-half to two-thirds scale reading and test for resistance faults.

> **REMEMBER:** Short the leads together every time a different range is used.

To identify a short, place the leads across the pair (*Fig. 1A*).

For measuring a ground, test the pair both sides to ground (*Fig. 1B*).

A cross is measured as a ground (*Fig. 1C*).

Cross battery also is measured as a ground, but will show battery during the DC test (*Fig. 1D*).

Multiply measurements by the range selected.

EXAMPLE:

100 measured in the X 1 K scale = 100,000 ohms.
    This scale measures accurately from 2 megohms to 20 megohms.

100 measured in the X 100 scale = 10,000 ohms.
    This scale measures accurately from 200,000 ohms to 2 megohms.

100 measured in the X 10 scale = 1000 ohms.
    This scale measures accurately from 20,000 ohms to 200,000 ohms.

100 measured in the X 1 scale = 100 ohms.
    This scale measures accurately from 0 ohms to 20,000 ohms.

Factors affecting ohms measurements

The distance to a fault is measured in ohms. When this resistance measurement is converted to its equivalent electrical footage, the technician must account for the following:

Gauge. The resistance of various conductors in

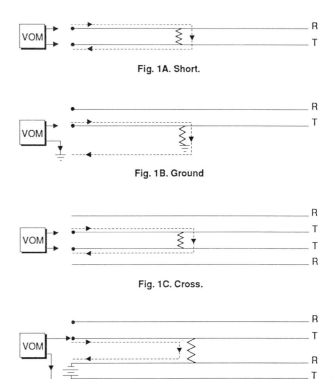

Fig. 1A. Short.

Fig. 1B. Ground

Fig. 1C. Cross.

Fig. 1D. Battery cross.

48VDC

-----►---- = DIRECTION OF CURRENT FLOW

Fig. 1. Types of Resistance Faults.

the cable route is determined by the diameter of the wire. Each gauge has a different ratio of ohms-to-feet.

Temperature. Heat increases resistance in a wire; cold lowers resistance. The temperature factor must be exactly determined for accurate measurements.

Wire composition. The makeup of the wire, whether copper, aluminum, or a mixed alloy, affects resistance and must be considered.

Helical cable design. Conductors are twisted into pairs, then into sub-units, units, and finally cable. As these helical twists vary conductor lengths, they must be accounted for when pinpointing where, beneath the cable sheath, the fault occurs.

Any of the above factors may change along a cable route. Each change must be accounted for or the electrical footage to the fault and the actual footage to the fault will be different.

On a resistance bridge, the gauge and temperature are entered into the test set, and the set automatically accounts for the helical design of the cable. Wire composition other than copper normally can be compensated for by using the gauge and temperature control to adjust for the characteristics of the wire.

Resistance bridge operation

Most resistance bridges in the field today use very similar principles to locate trouble. Some steps on the latest sets are done internally by microprocessor, but the principles of use are the same. When operating any resistance bridge, use the following step-by-step procedure:

Identify the faulted conductor. Use a high-impedance meter or the CO computer. If both sides of the pair are faulted, use the conductor that shows the heaviest fault.

Test the good pair. Use a high-impedance meter or the CO computer. The good pair must test good into the megohm range. Any fault on this good pair will cause an inaccurate fault measurement. (Careless testing of the good pair seems to give the field technicians the most grief in fault locating.)

> **CAUTION:** A low-impedance meter is unacceptable for testing the good pair. High-resistance faults above 3.5 megohms cannot be detected by the meter, and test measurements will be invalid. Additionally, the internal battery will dry out faults that could otherwise be located.

Many technicians are using meters that are incapable of testing resistance faults above 20 Megohms. This has no effect on the faulted conductor, but any higher resistance fault on the good pair will adversely affect the accuracy of a bridge measurement.

NOTE: *A separate "good pair" strung along the ground (Fig. 2) between accesses is recommended. This separate good pair can be any length, gauge or temperature, and it does not have to follow the actual cable path. Any extra wire can be left on the reel. Most new bridges in the field measure only the resistance of the faulted conductor.*

Attach the far-end strap. A resistance bridge requires a far-end strap. On a single pair fault, with one side testing good, simply strap across the pair and measure the distance to the fault (Fig. 3).

When measuring a short or when both sides of a pair are faulted, a separate good pair must be found. Strap both sides of the good pair to one side of the shorted pair or to the side of the pair which shows the heaviest trouble (Fig. 4).

> **REMEMBER:** In any measurement, it is always preferable to use a separate good pair.

Most modern resistance bridges have a built-in ohmmeter which shows a valid strap. If a valid strap is not indicated by the meter, the pair is either open or transposed. Excessive resistance indicates the faulted conductor is "going open" and invalid readings will result. In this case, find another faulted conductor for measurements.

> **REMEMBER:** Use only an approved strap when shorting the pair at the far end. When strapping in ready access plant, use an approved connector. Do **not** skin the insulation from the conductors and twist the pairs together as this will add series resistance to the strap-to-fault measurement.

In a fixed count terminal where binding posts are present, use a braided-wire three-way strap. Do not use tinsel-type wire such as a "B" transfer clip. Such wire adds as much as two ohms of series resistance to the strap-to-fault measurement.

Gauge and temperature setting. When entering the gauge with a CO computer, use the gauge at the strap. With a field bridge and where more than one gauge is encountered, use the gauge at the test set. If the fault is near the strap, the

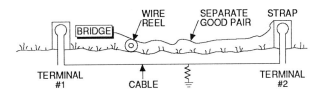

Fig. 2. Separate good pair laid on ground.

Fig. 3. One pair hookup.

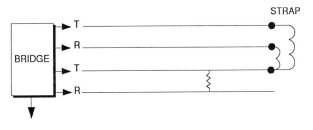

Fig. 4. Separate good pair hookup for short.

gauge can be changed to accommodate the strap-to-fault measurement.

Set the temperature control to the exact temperature of the conductor. If a temperature change is encountered, it must be treated as you would a gauge change.

Measure and record distance from test set to strap. This step is required for field bridges only. Any electrical measurement which differs from actual distance indicates:

■ A gauge change;
■ A slack loop;
■ A load coil;
■ Invalid cable map;
■ A faulty test set (check with self-test).

NOTE: *When measuring several faulted conductors in a cable, the footage readings will vary slightly due to the different twist factor on various pairs.*

Gauge change. This occurs when two or more gauges are spliced in the section. If a larger gauge is added, the test set will indicate less than the actual footage. For example, if a 300 foot section of 24 gauge cable had 50 feet of 22 gauge inserted at a cable cut, the section would measure 281 feet with a resistance bridge (*Fig. 5*).

If a smaller gauge is added, the test set will indicate more than the actual footage. In the above example, if 50 feet of 26 gauge cable had been used instead of 22 gauge, the resistance bridge would indicate 330 feet of cable in the section (*Fig. 6*).

Slack loops and butt splices. A slack loop or a butt splice will add footage to the section length (*Fig. 7*).

An extreme example of this type of underground anomaly occurred in Los Angeles, California. A technician called and said that his resistance bridge measured 70 feet to a cable fault. When the cable was exposed and no damage found, he opened the cable and remeasured the fault. The bridge showed 140 feet of cable in the section and still 70 feet to the fault. The bridge passed a self-test. From this evidence we suspected a slack loop. Fur-

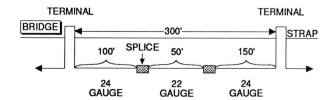

Fig. 5. DTS 281'.

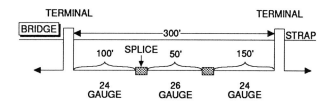

Fig. 6. DTS 330'.

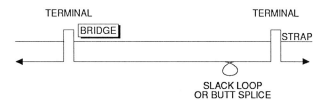

Fig. 7. Slack loop or butt splice adds to footage.

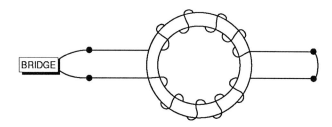

Fig. 8. Load coils add electrical distance.

ther investigation revealed 70 feet of 900 pair cable, still on its 415 reel, buried below the control point.

(We must assume that the construction crew which installed the cable figured that if the control point were, say, hit by a car and damaged, the repair group could simply unwrap another loop of cable and re-splice the box.)

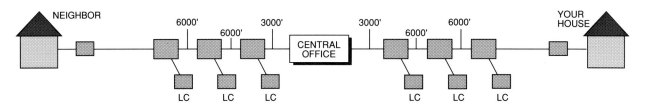

Fig. 9. Circuit loading.

Loading coils. A load coil adds approximately four ohms to a conductor measurement (*Fig. 8*). This must be accounted for when measuring.

Use the following chart to determine equivalent footage for each gauge:

At 68 degrees:

19 ga.: 4 ohms × 124.24 ft. per each ohm = 497 ft.

22 ga.: 4 ohms × 61.75 ft. per each ohm = 247 ft.

24 ga.: 4 ohms × 38.54 ft. per each ohm = 154 ft.

26 ga.: 4 ohms × 24.00 ft. per each ohm = 96 ft.

NOTE: *Different types of load coils made by different manufacturers will vary in resistance from about 3.6 to 4.2 ohms. The above chart is an average.*

REMEMBER: Four ohms in a load is for each conductor, not the pair. When measuring a solid short, use eight ohms.

EXAMPLE: A load coil is in a 1000-foot section of 24 gauge cable. With the gauge switch set to 24 and the proper temperature entered, the bridge indicates 1150 feet of cable. The extra 150 feet is the approximately four ohms of added resistance on one conductor by the load coil.

Load coils are designed to offset the natural capacitive attenuation of a POTS (Plain Old Telephone Service) cable pair. The one milihenry of natural inductance per mile does not offset the .083 micro farads capacitance per mile of the cable itself. Load coils (inductors) are used in any circuit longer than 18,000 feet, and are added first at 3000 feet from the CO, and then at 6000 feet spacing for the remainder of the circuit. (The last 10,000 feet to the subscriber does not require loading.)

The 3000-foot spacing from the CO is to balance the loading. If you call your neighbor, the load coil on your pair is at 3000 feet from the CO and the load coil on his cable pair is 3000 feet from the CO. This gives the total circuit (yours and his) the necessary 6000-foot spacing between load coils (*Fig. 9*).

NULLING: clearing and balancing the bridge. Most field bridges have a NULL 1 or a ZERO circuit which, when centered, removes the effect of spurious AC or DC voltage that may be on the circuit.

If NULL 1 or ZERO drifts back and forth, the pair selected for a good pair is faulted. This can be remedied by using another good pair in the cable or an external separate good pair.

The BALANCE or NULL 2 control balances an internal model of the test circuit at the fault. When the meter is centered again, the set stores the proportional resistance to be interpreted as distance-to-fault and strap-to-fault.

If NULL 2 or BALANCE drifts erratically, any readings are invalid. Expect more than one fault on the bad conductor, and more than just a bad splice. Suspect water in a section.

Distance-to-fault. Portable bridges measure the distance to the faulted conductor from the test set. If a gauge change, temperature change or any other resistance change occurs in the fault conductor, it must be accounted for in the measurement.

Strap-to-fault. The strap-to-fault measurement reads from the strap back to the fault. The same rules apply to strap-to-fault as to distance-to-fault.

When using the bridge, measure enough conductors to dig confidently without the aid of an earth-gradient frame. Use graph paper and record all measurements until a specific pattern is formed.

When performing section analysis, series resistance (the pair going open) on the faulted conductor indicates that the fault will be eventually found in a splice or an encapsulation. These series resistance faults normally occur on the tip side of the cable pair in POTS.

REMEMBER: Faulty test results are *always* the fault of the test set and never the fault of the inexperienced operator using improper test procedures.

Adjusting measurement for resistance factors

Adjusting for gauge: The feet-per-ohm in the most common gauges of conductors at 68 degrees Fahrenheit are as follows:

19 gauge	124.24	feet/ohm
22 gauge	61.75	feet/ohm
24 gauge	38.54	feet/ohm
26 gauge	24.00	feet/ohm

EXAMPLE: The CO computer reports a test on a 150 k ohm short with a strap-to-fault measurement of 10 ohms in 24 gauge cable at 68 degrees. Using the above chart, the technician calculates:

10 (ohms) × 38.54 (ft/ohm) = 385.4 feet

The technician will find the fault 385 feet from the strap. As long as there is no gauge change, the information would be accurate.

Adjusting for temperature: The computer operator enters and computes the temperature factor at the time of the measurement. If the temperature entered is wrong, each degree of error from the actual conductor temperature will create .00218 feet of error per foot of wire.

■ As the temperature of the wire increases, so does its resistance, and the number of feet-per-ohm will decrease.

■ As the temperature decreases, so does resistance and the number of feet-per-ohm increases.

As our chart above is calculated at 68 degrees, the difference in temperature from 68 degrees is used in the formula.

EXAMPLE: The CO computer reports a test on a 150 k ohm short with a strap-to-fault measurement of 10 ohms in 24 gauge cable at 68 degrees. As we have seen, his original calculation shows the fault at 385 feet.

But in the field, the technician finds actual conductor temperature at 58 degrees instead of 68 degrees. To compute the effect of this change on the measurment, use the following formula:

Distance @ 68 × **error/degree** × **diff. in temp** = **difference**

$$385 \times .00218 \times 10 \text{ degrees} = 8.393$$

As the resistance has *dimished* due to *lower* temperature, *add* this footage to the original measurement:

$$385 + 8.393 = 393.393 \text{ feet strap-to-fault}$$

> **REMEMBER:** As the temperature decreases, ADD the difference; as the temperature increases, SUBTRACT the difference.

EXAMPLE 2: A bridge measurement indicates a fault 40 ohms from the strap. The cable map indicates 22 gauge cable. The conductor temperature is 90 degrees. Convert 40 ohms of 22 gauge to its electrical equivalent at 68 degrees:

$$40 \text{ (ohms)} \times 61.75 = 2470 \text{ feet}$$

Subtract 68 degrees from 90 degrees. This equals a 22 degree difference.

Using our formula:

Distance @ 68 × **error/degree** × **diff. in temp** = **difference**

$$2470 \times .00218 \times 22 = 118.5$$

As the higher temperature has added

resistance, *subtract* the difference from the original distance: 2470 feet − 118.5 feet = 2351.5 feet to the fault at 90 degrees.

> **REMEMBER:** Buried conductor temperature can be as low as 28 degrees in the frost and up to 120 degrees under asphalt. Aerial conductor temperature can vary from 40 degrees below zero to 160 above.

A thermometer under tap water will indicate buried temperature in many instances. Aerial cable will be at ambient temperature when cloudy or shaded and increase to as much as 40 degrees above ambient temperature in the hot sun.

A temperature change is calculated and treated much the same as a gauge change. A measurement through cable part aerial at 130 degrees and part buried at 75 degrees must consider and calculate the effect of temperature on both sections.

> **REMEMBER:** When converting ohms to feet, if any of the above parameters change, they must be factored into the calculations.

Adjusting for cable composition: Most cable conductors are of copper, although aluminum was used for several years. Aluminum has a higher resistance and a different number of feet per ohm. For example, 22 gauge copper equals 61.75 feet per ohm at 68 degrees, and 22 gauge aluminum equals 37.08 feet per ohm at that temperature.

"B" service wire, used by the Bell Operating Companies, is an alloy of copper and steel. Its diameter is 20 gauge, but to convert ohms to feet you must compensate for the steel alloy with the gauge and temperature controls on the bridge. Set in 24 gauge and lower the temperature by 26 degrees from ground temperature.

Accounting for helical cable design: When cable is manufactured, a twist is put in the pair, and a lay in the sub-units and units, and in the cable. Any conductor measurement (compared to sheath-length) would be long if the helical design is not accounted for. Since different pairs have different twist factors, each pair will measure differently.

For example, an untwisted white/blue pair (2.0 Inches per twist) in 100 feet of cable, would be 103 feet long. The red/slate pair (4.7 inches per twist) would only be 101 feet.

As an average, subtract 2% from the conductor length to calculate sheath length.

In the real world, these situations are encountered almost daily. For an in-depth study of techniques for fault locating through multiple changes in resistive characteristics, we move to gauge changes.

REVIEW QUESTIONS:

1. How is current flow measured?
A. Connect the VOM in series.
B. Use the DC probe.

2. For measurement in a fixed-count terminal use:
A. A "B" transfer clip.
B. A braided wire three-way strap.
C. A needle-noise pliers.

3. If a large gauge cable is added to a section already containing a smaller gauge cable, the test set footage indication will be:
A. More than actual footage.
B. Less than actual footage.

4. When converting an ohms measurement to footage, the technician must account for:
A. Gauge.
B. Temperature.
C. Wire composition.
D. Helical cable design.
E. All of the above.

5. When using a separate good pair strung along the ground, the good pair must:
A. Follow the exact cable route.
B. Be the same gauge and temperature as the cable being tested.
C. Be a larger gauge than the cable being tested.
D. Be strapped to the faulted conductor at the far end.

(Answers on page 89)

CHAPTER 5

OUTLINE

Gauge changes

Examples of gauge conversion
- Single gauge change
- Multiple gauge changes and load coils
- Temperature change
- Accounting for load coil

Sample problems

OBJECTIVES

After completing this chapter, the student will be able to:

1. Account for changes in the gauge of cables whether or not they are indicated on a cable map.

2. Account for temperature changes and the presence of load coils.

Gauge Conversion

As you read, watch for the answers to the following important questions:

1. Why is it important to perform gauge conversion?

2. What factors should be considered in locating a suspected gauge change?

3. What is the impact of temperature change for every foot of cable?

As we've seen, any change in resistance along a conductor route will affect bridge measurements. Such resistance changes are caused by:

- Gauge changes
- Temperature changes
- Loading coils

Gauge changes

When a resistance bridge is testing through a gauge change, the restive difference between the two gauges must be accounted for to determine physical distance to the fault.

Ideally, bridge readings should be compared to cable maps, and the lengths of the various gauges in the route converted to ohms or the equivalent electrical footage. This would give an accurate electrical map of the test area. Unfortunately, field technicians often encounter inaccurate, outdated cable maps. When the principles of gauge conversion are fully understood, such map problems can be circumvented using Electronic Reference and Bridge Theory (Chapter 6).

The following chart is used to convert from one gauge to another:

TO CONVERT	TO	MULTIPLY BY
19 gauge	22 gauge	0.497
	24 gauge	0.31
	26 gauge	0.193
22 gauge	19 gauge	2.01
	24 gauge	0.624
	26 gauge	0.389
24 gauge	19 gauge	3.22
	22 gauge	1.6
	26 gauge	0.623
26 gauge	19 gauge	5.18
	22 gauge	2.57
	24 gauge	1.61

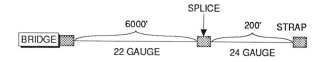

Fig. 1. Single gauge change.

Examples of gauge conversion

Single gauge change. In *Fig. 1*, a strap-to-fault measurement is 300 feet when measured in 24 gauge at 75 degrees.

■ The cable map indicates that only 200 feet of cable is 24 gauge at 75 degrees and the rest of the cable is 22 gauge at 75 degrees.

Measurement: 300 ft. of 24 ga. at 75°
Subtract: −200 ft. of 24 ga. at 75°

Leaving: 100 ft. of 24 ga. at 75° to be converted
to 22 ga. at 75°.

Using the conversion factor from the chart:

Original footage × factor = converted footage
100 (24 ga.)   ×   1.6   =   160 (22 ga.)

To locate the fault, add the 200 feet of 24 gauge to the 160 feet of 22 gauge. The fault is 360 physical feet from the strap.

Multiple gauge changes and load coils. *Fig. 2* shows a test area with several gauge changes plus a load coil.

The central office bridge, testing 24 gauge at 60 degrees gets a strap-to-fault measurement of 1621 feet.

■ The map indicates that only the first 115 feet of cable from the strap is 24 gauge at 60 degrees.

Measurement: 1621 ft. of 24 ga. at 60°
Subtract: −115 ft. of 24 ga. at 60°

Leaving: 1506 ft. of 24 ga. at 60°
(Figures are rounded to nearest foot)

■ The map indicates that the next section is 470 feet of 26 gauge. The 470 feet of 26 gauge at 60 degrees must be converted to 24 gauge and then subtracted from the remaining 1506 feet of 24 gauge at 60 degrees.

To convert 470 feet of 26 gauge to 24 gauge, use the formula:

Original footage × factor = converted footage
470 (26 ga.)   ×   1.61   =   757 (24 ga.)

Therefore:

Measurement: 1506 ft. of 24 ga. at 60°
Subtract: −757 ft. of 24 ga. at 60°

Leaving: 749 ft. of 24 ga. at 60°

■ The map indicates that there is a load coil in series with the section and the approximately four ohms of load coil resistance must be converted to 24 gauge at 60 degrees. Use the following chart for load coils:

19 ga.: 4 ohms × 124.24 ft. per each ohm = 497 ft.

22 ga.: 4 ohms × 61.75 ft. per each ohm = 247 ft.

24 ga.: 4 ohms × 38.54 ft. per each ohm = 154 ft.

26 ga.: 4 ohms × 24.00 ft. per each ohm = 96 ft.

(There is a minimal temperature change in the load measurements. We will not account for it in this exercise.)

Measurement: 749 ft. of 24 ga. at 60°
Subtract: −154 ft. of 24 ga. load coil

Leaving: 595 ft. of 24 ga. at 60°

■ The map indicates the next section is 460 feet of 19 gauge at 60 degrees. To convert:

Original footage × factor = converted footage
460 (19 ga.)   ×   .31   =   143 (24 ga.)

Measurement: 595 ft. of 24 ga. at 60°
Subtract: −143 ft. of 24 ga. at 60°

Leaving: 452 ft. of 24 ga. at 60°

■ The map indicates the next section is 200 feet of 22 gauge at 60 degrees. To convert:

Original footage × factor = converted footage
200 (22 ga.)   ×   .624   =   125 (24 ga.)

Measurement: 452 ft. of 24 ga. at 60°
Subtract: −125 ft. of 24 ga. at 60°

Leaving: 327 ft. of 24 ga. at 60°

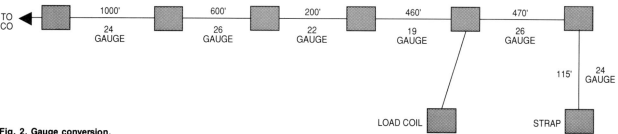

Fig. 2. Gauge conversion.

■ The next section indicates 600 feet of 26 gauge at 60 degrees. Convert:

Original footage × **factor** = **converted footage**
600 (26 ga.)   ×   1.61   =   966 (24 ga.)

Measurement:	327 ft. of 24 ga. at 60°
Subtract:	−966 ft. of 24 ga. at 60°
Leaving:	A negative number

This indicates the fault is in the 26 gauge section. To find how far into the section, convert the remaining 327 feet of 24 gauge to 26 gauge.

Original footage × **factor** = **converted footage**
327 (24 ga.)   ×   .623   =   204 (26 ga.)

The fault is into the 26 gauge section 207 feet from the 22 gauge splice.

Temperature change. *Fig. 3* shows an aerial cable in the sun with an outside temperature of 90 degrees. Measurements are taken at 130 degrees to account for the internal conductor temperature. Distance-to-fault measures 1500 feet of 24 gauge at 130 degrees. The map indicates only 1000 feet of cable at 130 degrees. The cable then goes buried back to a manhole 3000 feet away. The buried temperature is 70 degrees.

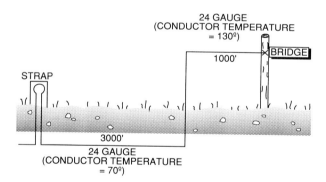

Fig. 3. DTF = 1500′.

To calculate the distance from where the cable goes buried, use the following technique:

Each degree of temperature change will affect the measurement by .00218 feet for each foot of cable.

Measurement:	1500 ft. of 24 ga. at 130°
Subtract:	−1000 ft. of 24 ga. at 130°
Leaving:	500 ft. of 24 ga. at 130°
	to be converted to 70°

     130 degrees
   − 70 degrees
     = 60 degrees of temp. change.

Using the formula:

Temperature Change × **ft/degree** = **change/ft**
60 degrees   ×   .00218   =   .1308

Multiply the change/foot by the section footage:

.1308 x 500 feet = 65.4 feet difference.

When the temperature change decreases, the difference must be added.

500 feet + 65.4 feet = 565.4 feet.

The fault is 565.4 feet into the buried section or 1565.4 feet from the test set.

Sample problems:

Using the above information, calculate the actual distance or strap-to-fault in the following examples.

1. (*Fig. 4*). **STF** =700 feet of 22 gauge at 50 degrees. The map indicates that from the strap, there is 300 feet of 22 gauge and the remaining 6000 feet is 24 gauge.

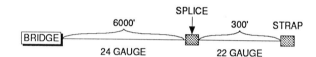

Fig. 4. STF = 700′.

2. (*Fig. 5*). **DTF** =2000 feet of 24 gauge at 130 degrees. The map indicates 1500 feet of 24 gauge at 130 degrees. At the point where the cable goes buried, the map shows 6000 feet of 22 gauge and the temperature changes to 60 degrees.

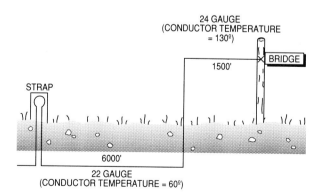

Fig. 5. DTF = 2000′.

3. (*Fig. 6*). **DTF** =300 feet of 26 gauge at 55 degrees. The map indicates 140 feet of 26 gauge, then 1500 feet of 24 gauge at 55 degrees.

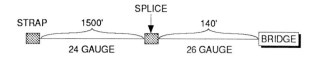

Fig. 6. DTF = 300'.

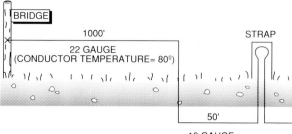

Fig. 7. STF = 450'.

4. (*Fig. 7*). **STF** =450 feet of 19 gauge at 60 degrees. The map indicates 50 feet of 19 gauge cable at 60 degrees. the cable goes aerial to 80 degrees and changes to 1000 feet of 22 gauge.

Answers:

1. Measurement: 700 ft. of 22 ga. at 50°
    Subtract: −300 ft. of 22 ga. at 50°

    Leaving: =400 ft. of 22 ga. reading to be converted to 24 ga. at 50°.

Original footage	×	**factor**	=	**converted footage**
400		.624		249.6 feet

    Therefore: 249.6 ft. (24 ga. section)
              +300 ft. (22 ga. section)
              =549.6 ft. STF.

2. Measurement: 2000 ft. of 24 ga. at 130°
    Subtract: −1500 ft. of 24 ga. at 130°

    Leaving: 500 ft. of 24 ga. reading at 130° to be converted to 22 ga. at 60°.

Gauge change formula:

Original footage	×	**factor**	=	**converted footage**
500		1.6		800

Temperature change formula:

Temperature change	×	**ft/degree**	=	**change/ft**
130 − 60 = 70		.00218		.1526

Multiply the change/foot by converted footage in 22 ga.:

$$.1526 \times 800 = 122.08$$

When temp change decreases, add the difference:

$$800 + 122.08 = 922.08$$

Therefore:
     1500 ft. (24 ga. section @ 130°)
     + 922.08 (22 ga. section @ 60°)
     =2422.08 feet DTF

3. Measurement: 300 ft. of 26 ga. at 55°
    Subtract: −140 ft. of 26 ga. at 55°

    Leaving: 160 ft. of 26 ga. reading at 55° to be converted to 24 ga. at 55°.

Original footage	×	**factor**	=	**converted footage**
160		1.61		257.6

Thus:
     257.6 (24 ga. section)
     +140 (26 ga. section)
     =397.6 feet DTF.

4. Measurement: 450 ft. of 19 ga. at 60°
    Subtract: − 50 ft. of 19 ga. at 60°

    Leaving: 400 ft. of 19 ga. at 60° to be converted to 22 ga. at 80°

Gauge change formula:

Original footage	×	**factor**	=	**converted footage**
400		.497		198.8

Temperature change formula:

Temperature change	×	**ft/degree**	=	**change/ft**
80 − 60 = 20		.00218		.0436

Multiply the change/foot by convered footage in 22 ga.:

$$.0436 \times 198.9 = 8.66768 \text{ ft.}$$

When temp change increases, subtract the difference:

$$198.8 − 8.66768 \text{ (9 ft.)} = 189.8 \text{ ft.}$$

Therefore:
     189.8 ft. (22 ga. section @ 80°)
     + 50 ft. (19 ga. section @ 60°)
     =239.8 ft. STF

For gauge conversion, ignore the footage of any laterals shown on the map. As a resistance bridge measures the path of least resistance, fault measurements do not include the resistance of a lateral. If the measurement reads to the vicinity of a lateral, however, there is a possibility that the fault is in the leg.

To locate in this type of plant, it is necessary to know techniques of bridge theory and electronic reference. We move to these subjects next.

FOR YOUR PERSONAL COMMENTS, OBSERVATIONS AND NOTES:

CHAPTER 6

OUTLINE

Resistance bridge theory
Common wire theory
Electronic reference

OBJECTIVES

After completing this chapter, the student will be able to:

1. Recognize what a resistance bridge can and can't do.

2. Handle a suspected lateral fault.

3. Use electronic reference techniques.

Strategies for Locating Faults

PREVIEW QUESTIONS

As you read, watch for the answers to the following important questions:

1. How is common wire theory used in fault locating?

2. What are the benefits of electronic reference techniques?

A thorough knowledge of bridge theory enables the technician to know what the bridge sees, what the bridge doesn't see, and how to efficiently use the set to quickly and accurately isolate and find the fault.

When testing between two accesses, a resistance bridge shows a fault on a lateral as in its three-way splice on the main cable. The distance of the fault from the tap does not matter, as the bridge reads the trouble as a fault on the main cable at the splice.

When reading a fault in the vicinity of the lateral spice, the fault is in one of three locations:

■ In the three-way splice;
■ On the main cable close to the leg;
■ Or somewhere down the leg.

Common wire theory

By moving either the bridge or the strap to the suspected lateral, the technician can prove the fault location without opening the three-way splice, unless that's where the fault has occurred. We call this the "common wire" method of fault locating. This common wire is the portion of the measurement which is included in the two tests.

When the strap is moved, the common wire is the Distance-To-Fault (DTF) measurement (*Fig. 1*).

When the set is moved, the common wire is the Strap-To-Fault (STF) measurement (*Fig. 2*).

As the bridge measures a single wire, not a pair, this common measurement allows a comparison between the readings which pinpoints the fault. The following examples demonstrate bridge theory using a portable bridge at 1400 Talbott Lane, 1400 Hamilton Road and 1400 Townes Drive.

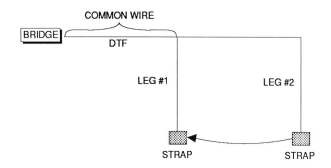

Fig. 1. Strap moved from leg #2 to leg #1.

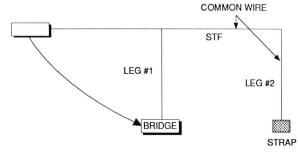

Fig. 2. Bridge moved to the end of leg #1.

EXAMPLE 1: Proving to a lateral.

In *Fig. 3*, a technician, dispatched on a fault in the 1-50 count, opens the pair at 1400 Talbott Lane. When testing the pair with a VOM, the fault shows beyond the cross-box either in any one of four laterals with the same 1-50 count or in the main cable.

The technician attaches a resistance bridge to the faulted conductor at 1400 Talbott Lane and straps the faulted conductor at the end of lateral #4. (Always strap at the farthest terminal to bracket the fault and work back to it.)

When set to the 24 gauge of the conductor and the correct 50 degrees conductor temperature, the bridge measures a Distance-to-Strap (DTS) of 3000 feet. The map indicates 3040 feet of cable, but the measurement is acceptable at this distance. (Temperature variations, mapped footage errors, gauge changes, and set accuracy could all account for the discrepancy). After nulling or balancing the bridge, the DTF is 1500 feet and the STF reading is 1500 feet.

These measurements indicate that the fault is either in leg #2 or in the vicinity of leg #2. To prove the location, the technician moves the strap to the end of leg #2 and measures the fault again (*Fig. 4*).

When placing the strap, the technician notes that the cable in lateral #2 is 22 gauge, but measures again in 24 gauge. The measurements with the strap at the end of leg #2 are:

DTS = 2000 feet of 24 gauge @ 50 degrees
DTF = 1600 feet of 24 gauge @ 50 degrees
STF = 400 feet of 24 gauge @ 50 degrees

Analysis of the measurements are:

DTS = 2000 feet: This measurement is 24 gauge conductor from the test set to the three-way splice, and through 22 gauge resistance

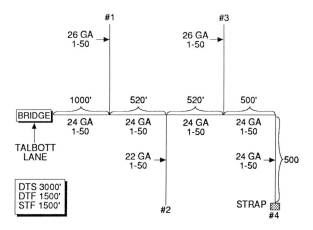

Fig. 3. Always strap at farthest terminal.

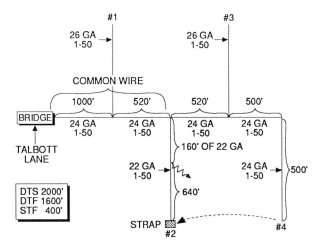

Fig. 4. Move strap and remeasure.

from the three-way to the strap at the end of leg #2 (measured as 24 gauge).

DTF = 1600 feet: This measurement is 1500 feet of 24 gauge to the three-way splice plus 100 feet of 24 gauge resistance that, when converted to 22 gauge (100 × 1.6 = 160 feet), indicates the fault 160 feet from the three-way splice down leg #2.

STF = 400 feet of 24 gauge measurement to be converted to 22 gauge (400 x 1.6 = 640 feet). The fault is 640 feet from the strap.

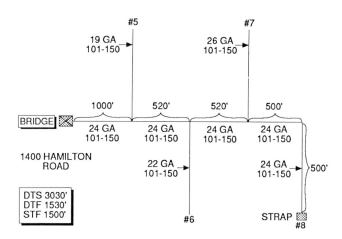

Fig. 5. Fault proved to vicinity of leg #6.

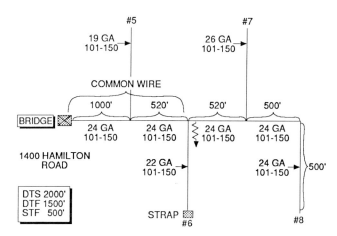

Fig. 6. Measurements with strap moved to leg #6.

Note that the cable map does not indicate the length of the lateral. To determine the length of leg #2, add the 160 feet from the three-way splice to the fault in leg #2 and the 640 feet from the end of leg #2 to the fault. Lateral #2 is 800 feet long.

When analyzing bridge readings, identify which wire is used for both measurements. In this example, DTF was used because all of the conductor from the cross-box up to leg #2 was used both times. If the technician had strapped the pair in the cross-box and moved the test set from lateral #4 to lateral #2, then STF measurements would have been used.

This reading is referred to as the Common Wire Reading (CWR).

> **RULE:** When using a resistance bridge to determine whether a fault is in a lateral, three-way splice, or on the main cable, if the Common Wire Reading (CWR) *increases* when the strap is moved to the lateral, the fault is in the lateral.

EXAMPLE 2: Proving to the cable beyond the lateral.

In *Fig. 5*, a technician is dispatched to a resistance fault in the 101-150 count. When the pair is opened at 1400 Hamilton Road, the fault is shown as beyond the cross-box in the distribution cable. The resistance bridge is attached at the cross-box, and the strap is placed at the end of the run at leg #8.

With 24 gauge and the correct temperature set, the resistance bridge reads:

DTS = 3030 feet
DTF = 1530 feet
STF = 1500 feet

Analysis of these measurements indicates the fault in the vicinity of lateral #6. Move the strap to the end of lateral #6, (*Fig. 6*).

With the test set still at 1400 Hamilton Road and the strap at the end of leg #6 the readings are as follows:

DTS = 2000 feet of 24 gauge @ 50 degrees
DTF = 1500 feet of 24 gauge @ 50 degrees
STF = 500 feet of 24 gauge @ 50 degrees

Analysis of the resistance measurements are as follows:

DTS = 2000 feet: This measurement is 24 gauge conductor from the test set to the three-way splice, and through 22 gauge resistance from the three-way to the strap (measured as 24 gauge.)

DTF = 1500 feet. This is the electrical distance from the test set to the three-way splice. Note that the strap at leg #6 now shows as a fault in the three-way (1500 feet). As the original DTF reading (strapped at leg #8) was 1530 feet, and the wire common to the two readings is from the test set to the three-way, the fault is proved 30 feet *beyond* the leg #6 three-way.

STF = 500 feet. This is the electrical length of leg #6 when read as 24 gauge. To find the actual length of leg #6, multiply 500 × 1.6 = 800 feet.

> **RULE:** When using a resistance bridge to determine whether a fault is in a lateral, three-way splice, or on the main cable, if the Common Wire Reading (CWR) *decreases* when the strap is moved, the fault is *beyond* the lateral in the main cable, because the strap now produces a fault at the tap. The difference between the two common wire readings is the distance on the main cable beyond the splice.

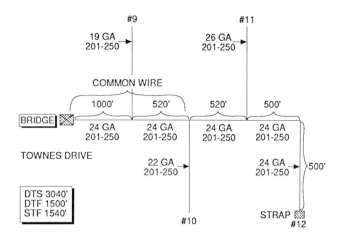

Fig. 7. Fault proved to vicinity of leg #10.

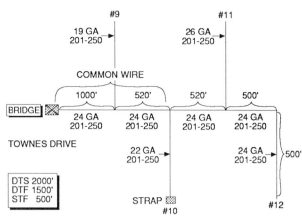

Fig. 8. Strap moved to leg #10.

EXAMPLE 3: Proving to the splice or before the splice.

In *Fig. 7*, a technician is dispatched to a fault in the 201-250 count. When the pair is opened at 1400 Townes Drive, the VOM indicates the fault in the distribution plant. The technician connects the bridge to the faulted conductor at the cross-box and straps the pair at the end of lateral #12.

After setting proper gauge and temperature, the test set reads:

DTS = 3040 feet of 24 gauge @ 50 degrees
DTF = 1500 feet of 24 gauge @ 50 degrees
STF = 1540 feet of 24 gauge @ 50 degrees

The measurements indicate the fault in the vicinity of lateral #10.

With the strap at the end of leg #10 (*Fig. 8*), the measurements are as follows:

DTS = 2000 feet of 24 gauge @ 50 degrees
DTF = 1500 feet of 24 gauge @ 50 degrees
STF = 500 feet of 24 gauge @ 50 degrees

Analysis of the resistance measurements are as follows:

DTS = 2000 feet: This measurement is 24 gauge conductor from the test set to the three-way splice, and through 22 gauge resistance from the three-way to the strap at the end of leg #10 (measured as 24 gauge).

DTF = 1500 feet. As this measurement is the same as when strapped at leg #12, the fault is not in the lateral, but on the common wire, either in the three-way at splice #10 or on the cable between lateral #10 and lateral #9. One more measurement is necessary to locate the fault (*Fig. 9*).

Move the test set to the end of leg #12 leaving the strap at lateral #10. Note that strap-to-fault is now the CWR.

If this CWR is the same 500 feet as the reading when strapped at leg #10, the fault is in the three-way splice.

If the CWR decreases, for example from 500 feet to 470 feet (*Fig. 10*), the fault is 30 ft.

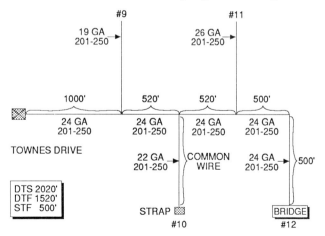

Fig. 9. Move test set to leg #12.

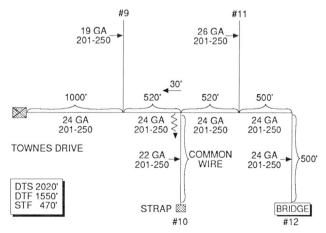

Fig. 10. CWR decreases. Fault is 30' to left of leg #10.

to the left of leg #11, as the set is now reading to the splice, not the fault.

> **RULE:** If the CWR *stays the same* when the test set is moved, the fault is in the three-way splice. If the CRW *decreases*, the fault is beyond the splice, reference the test set. In this case, the fault is between leg #9 and #10.

Electronic reference

Most new construction uses the Serving Area Concept (SAC) where the customer drop is spliced directly into the cable with pairs dedicated to the customer.

In older areas with ready access plant, most companies are locking up the terminals

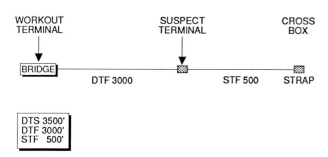

Fig. 11. Electronic reference measurement #1.

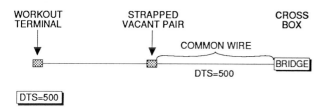

Fig. 12. Fault confirmed in terminal.

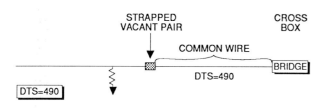

Fig. 13. Fault confirmed 10 feet beyond terminal.

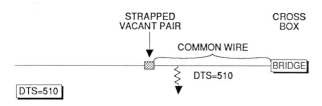

Fig. 14. Fault confirmed 10 feet toward cross-box.

to prevent unnecessary access to the cable pairs. Many of these terminals multiple in several places.

Because of this reduced access, frustrated technicians are in the habit of entering locked up plant or removing the "U" guard on a telephone pole and removing a 10-12 inch piece of sheath to gain access to cable pairs.

Through a good understanding of bridge theory, and by using Electronic Reference, such unnecessary entries are eliminated, and the quality of the plant will be much improved.

The following is an example of Electronic Reference techniques:

A technician attaches the bridge at the work-out terminal, straps the pair at the cross-box and measures the distance to a faulted conductor with the following results (*Fig. 11*).

DTS = 3500 feet of 24 gauge @ 75 degrees
DTF = 3000 feet of 24 gauge @ 75 degrees
STF = 500 feet of 24 gauge @ 75 degrees

Using the cable map, the technician finds a terminal at approximately the indicated footage. A visual examination of the terminal shows no obvious fault and it is locked up. The measured pair is not accessible unless it is opened.

Before opening the terminal, the following tests will prove the trouble:

Strap a vacant pair that appears at that terminal and measure the DTS from the cross-box (*Fig. 12*).

If the DTS is 500 feet of 24 gauge at 75 degrees, this agrees with the original STF measurement and confirms the fault in the terminal (*Fig. 13*). Open and fix it.

If the DTS is *less* than the original 500 feet (*Fig. 13*), the fault is *beyond* the terminal the difference between the readings. This is because the strap places a short on the common wire *before* the fault (*Fig. 14*).

If the DTS is *more* than the original 500 feet, the fault is *toward* the cross-box the difference between the readings. Here, the reading to the strap at the terminal is more than the measured distance to the actual fault.

Distances to cross-boxes, splices, and terminals should be measured for future benchmarks. This decreases the time needed to localize cable faults in the future.

TERMS TO REMEMBER: *(Write the definitions in your own words.)*

Common Wire Reading (CWR)—...

....

Distance-to-Fault (DTF)—..

....

Strap-to-Fault (STF)—...

....

Serving Area Concept (SAC)—..

....

REVIEW QUESTIONS:

1. *A resistance bridge measures:*

A. *A single wire.*

B. *A pair.*

C. *Both of the above.*

2. *If the Common Wire Reading (CWR) increases when the strap is moved to lateral, the fault is in the:*

A. *Main cable.*

B. *Lateral.*

C. *Three-way splice.*

3. *If the fault is beyond the lateral and the strap is moved to the lateral, the CWR:*

A. *Increases.*

B. *Decreases.*

C. *Stays the same.*

4. *A resistance bridge measures all wire, including any laterals.*

A. *True.*   B. *False.*

5. *A gauge change will not affect resistance measurements.*

A. *True.*   B. *False.*

(Answers on page 89)

FOR YOUR PERSONAL COMMENTS, OBSERVATIONS AND NOTES:

CHAPTER *7*

OUTLINE

Capacitance and capacitors

Use of capacitance for measurements

Built-in capacitance

Capacitance troubles

Using the two-terminal open meter

How to use the open meter
- Test the pair
- Ground the test set
- Measuring nonworking pairs
- Measuring unbalanced pairs
- Filled cable
- Special cables
- Water

Using a Time Domain Reflectometer (TDR)

Three-terminal open measurements

OBJECTIVES

After completing this chapter, the student will be able to:

1. Use capacitance for several measurements.

2. Cite factors affecting capacitance.

3. Operate an open meter.

4. Explain the uses of a Time Domain Reflectometer (TDR).

Capacitance and Using an Open Meter

PREVIEW QUESTIONS

As you read, watch for the answers to the following important questions:

1. What are the characteristics of capacitance in a cable?

2. For what tests is an open meter most useful?

3. What are the uses and advantages of a TDR?

In a great mouthful, capacitance may be defined as that attribute of a system of conductors and dielectrics that permits the storage of electrically separated charges when potential differences exist between the conductors.

Like resistance, capacitance is a characteristic of electric potential and current flow along and between conductors. It has a major effect on transmission quality.

A capacitor is a device which stores an electrical charge. It consists of two conductive plates separated by a dielectric (*Fig. 1*). The larger the plates and the thinner the dielectric, the greater the capacitance (storage).

A capacitor charges (with negative electrons flowing to the positive plate) until its potential equals the voltage of the charging battery (*Fig. 2a*). With the battery removed, the capacitor holds its charge (*Fig. 2b*). When the plates are reconnected by a conductor, current flows as a discharge until the voltage on the plates is equal (*Fig. 2c*).

Think of telephone plant as a giant capacitor. When a VOM (on the R X 1 k ohm scale) is attached across a pair or from either side to ground, the meter kick shows the extent of charge placed on the conductors by the meter battery. Once the charge equals the battery voltage, the needle drops off to zero. Remove the VOM and the conductor retains the charge until a path to ground is established.

Capacitance tests show:

■ The approximate length of a balanced pair.

■ The approximate distance to an open on an unbalanced pair.

■ Added capacitance when crossed with another nonworking pair.

■ The difference in length when the pair is

Fig. 1. Capacitor construction.

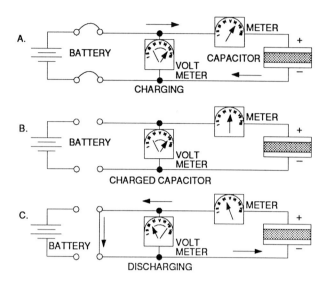

Fig. 2. Charging and discharging a capacitor.

open one side on a lateral or beyond the subscriber. This imbalance causes excessive noise.

Though useful as a measurement device, capacitance in telephone plant must be carefully controlled to assure transmission quality. Increases in mutual capacitance due to length of cable pair reduce the voice signal level. This affects high frequencies more than low frequencies (which is why a woman's voice might not be heard while a man's voice has no problems). To offset this attenuation, load coils are necessary in any circuits longer than 18,000 feet. They restore the signal level to proper amplitude.

Mutual capacitance is the sum of all cable capacitances acting on one pair (*Fig. 3*). This includes tip-to-ring, tip-to-shield, ring-to-shield, tip to all surrounding conductors, and ring to all surrounding conductors of the same unit or sub-unit (under one binder tie). The lay in each unit capacitively separates all units.

Tip-to-shield and ring-to-shield capacitance must be balanced to remove the effect of any induced AC. Induced voltage of one phase entering the ring conductor from the shield is canceled by the same voltage reversing its polarity and exiting on the tip. If the ring- or tip-to-shield capacitance is greater, as when one side of a pair is open on a lateral, the increase in uncanceled induced voltage is heard by the subscriber as audible noise.

The goal of manufacturers is a cable capacitance of .083 μF per mile. This can vary (within limits) however, and most open locates (especially before digging or cutting into the sheath) should be done by calibrating to the actual capacitance of a section with the test set.

Like resistance, capacitance is proportional to cable length and increases with distance. Thus, one mile of cable has a capacitance of .083 μF, two miles, .166 μF and so on.

Unlike resistance, capacitance is hardly affected by gauge or temperature. Gauge capacitance is controlled by the thickness of the

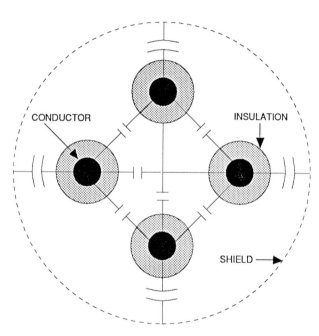

Fig. 3. Capacitance acting on a pair.

insulation at manufacture (*Fig. 4*), and temperature has little effect.

Cable capacitance is adversely affected by water intrusion (which increases it) and poor bonding (which decreases it).

Proper grounding increases the flow of current in the shielding circuit, improving noise performance of the cable. Shield continuity is essential to control capacitive balance for both transmission characteristics and safety.

The capacitance of the CO battery acts as a noise removing shunt filter to improve speech quality through an office.

Use of capacitance for measurements

A technician must have a working knowledge of capacitance to find open and split cable pairs in the field. The problem here is that experts do not agree on exactly how cable capacitance is used in troubleshooting.

Different measurement techniques are needed for nonworking cables and for working cables. Nonworking cables are normally tested for mutual capacitance (across the pair). Manufacturers test new cable during and after manufacture with this method, and it is used in testing new cable after it is spliced together, but before it is heated up.

Working cables are normally tested for ring or tip capacitance to ground. Fault locating technicians use these measurements after the cable is heated up and is working in the field.

This chapter involves working cables, and testing routines use capacitance-to-ground measurements. To use these measurements when testing in nonworking cables, we describe techniques which make a nonworking cable appear to be working cable to the test set.

Built-in capacitance

The manufacturer designs cable to a capacitance designated by the telephone company. Cables with an average mutual capacitance of .083 micro-farads per mile are used in POTS (Plain Old Telephone Service) circuits. Some special circuits, such as toll and carrier, may require a different mutual capacitance as higher frequency transmissions require lower capacitance.

Capacitance is a difficult entity to control and the manufacturer goes to great effort and expense to meet the required tolerances during cable manufacture.

In the field, modern splicing techniques, using good bonding and grounding, eliminate capacitance noise problems in a well-designed cable route. When splicing is complete, careful conformance testing assures a quality circuit.

Capacitance troubles

Splicing activity and inward and outward movement in the cable cause most open cable pairs. Distribution pairs are borrowed indiscriminately, adding to the problems. Mother Nature adds by electrolysis and galvanic corrosion of cable pairs. Bonds are left open and grounds are removed inadvertently.

When service is interrupted because of an open, normally the CO computer indicates whether the open is in the CO or the field. The computer shows the approximate location and whether the open is in the underground plant, the distribution plant, or the customer's drop or equipment. Unbalanced pairs (open beyond the subscriber, open on a lateral, or crossed with a nonworking pair) are displayed as opens by the office computer and must be retested and diagnosed in the field. Portable open meters, when programmed properly (and other resistive faults do not interfere with the measurements) will pinpoint the open.

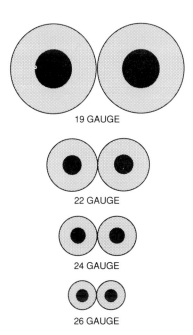

19 GAUGE

22 GAUGE

24 GAUGE

26 GAUGE

Fig. 4. Compensation for the gauge and distance so that all cables have the same capacitance.

> **REMEMBER:** Open meter measurements include all laterals, conductor beyond the subscriber, the capacitance of load coils, build-out circuits, any dead wire crossed with the pair, water, purging compound, or type of cable fill. When looking for an open cable pair, these capacitive anomalies affect any measurement.

Using the two-terminal open meter

The operation of most open meters is quite simple. With the meter connected to the open cable pair, enter the capacitance of the cable under test. Then measure and record the capacitive length of the ring and tip conductors and, if needed, their mutual capacitance. Successfully finding opens depends on the understanding of what will affect these measurements.

As the average mutual capacitance of an exchange cable pair is .083 μF/mile, this is the standard setting for most field and CO open meters used today. Using this average setting allows the technician to find the general vicinity of an open and may indicate in which splice or access it is. However, if the technician needs to dig a hole or open the cable sheath, this measurement is not reliable.

Mutual capacitance, as tested at the manufacturing plant, can be as high as .090 μF/mile and as low as .076 μF/mile and still be acceptable for customer use.

Measurements of an open in an .090 μF/mile cable with the meter set at the average .083 μF/mile are inaccurate. An open at 1050 feet in .090 μF/mile cable measures at 1200 feet with the open meter set to .083 μF/mile.

Measuring in a .076 μF/mile cable, an open at 1050 feet measures as 950 feet with meter set at .083 μF/mile.

Because of this variance, when digging and/or opening sheath, calibrate the set to the known length.

How to use the open meter

Test the pair. Use a VOM to test the conductors of the pair for resistance faults, DC voltage and induced AC voltage. When such troubles are excessive, an open meter reads long or gives invalid measurements.

Ground the test set. Make sure both the open meter and the cable section are properly grounded. To test for a good ground, measure any open cable conductor (tip or ring) to ground. Remove the test set ground and remeasure the conductor. The footage measurement should decrease by about one-third. If no change is noted, either the ground is bad or the ground cord on the set is open.

We encountered this situation in Oakland, California. A technician was measuring from a radio tower hut, which had an excellent ground when tested. The open cable pairs showed approximately 18,000 feet from the hut. Good pairs in the same count measured 20,000 feet. Because the fault was close to the far end, we moved to a BD box in the vicinity of 20,000 feet and measured back 1300 feet to the opens.

The same good conductors, which measured 20,000 feet from the other end, now measured about 13,000 feet from the box (a decrease of about one-third). When the ground was removed from the set, no change in

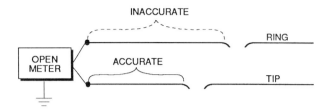

Fig. 5. Use the shorter reading.

measurement occurred. This confirmed a bad ground at the box. After grounding the box to a down-guy, the good pairs measured the correct 20,000 feet and the open conductors measured 2000 feet. Upon visual inspection, we found a bullet hole at this measurement.

Measuring unbalanced pairs. If the tip and ring measurements are different, the shorter of the two readings is to the open (*Fig.* 5). Because we are measuring capacitance to ground, the length of ungrounded conductor beyond the open changes the capacitance of the longer wire. Therefore, its total capacitance is affected, and any measurement along it is inaccurate.

> **REMEMBER:** Always use the shorter of two differing measurements to find the open.

Measuring nonworking pairs. There are two ways of measuring opens in nonworking units: ground 12 pairs in the unit; or calibrate to the known length on a good pair.

Most field open meters are designed to be used on working cables. Most operating manuals state that 12 or more pairs from the same unit or sub-unit under test must either be working or grounded to the shield for accurate measurements.

When measuring a balanced open cable pair, if the tip and ring measurements are shorter than the mutual capacitance measurement, the group is nonworking. Twelve or more pairs must be grounded to simulate a working group (*Fig.* 6).

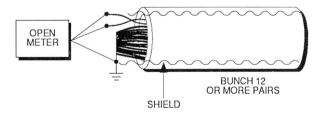

Fig. 6. Testing in non-working cable.

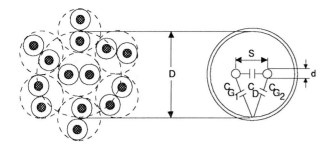

Fig. 7. The pair is surrounded by other pairs to form an electrostatic shield of effective diameter D.

For example, when measuring an open cable pair in the red/orange group of a 200 pair PIC cable with no working pairs, even if all other cable groups are working, this nonworking group needs 12 pairs grounded to the shield for an accurate measurement.

If this grounding is not done, nonworking pairs surrounding the test pair in a sub-unit or unit build an electrostatic shield which is closer than the actual cable shield. This electrostatic shield (consisting of capacitors added in series), causes the readings to be short (*Fig. 7*).

For example, in a nonworking group, a cable pair open at 420 feet will measure approximately 385 feet in the tip- or ring-to-ground measuring position. The mutual capacitance measurement will be approximately 405 feet.

If it is not possible to ground 12 pairs and the length of the test section is known, calibrate the open meter to that length on a good pair in the test group. When measuring in another group, the set must be recalibrated using a good pair in that group.

Filled cable. Filling compound is added to cables to stop the intrusion of water. The compound has a greater dielectric constant than the air in nonfilled cable, so mutual capacitance is increased.

The manufacturers first used a proportionately thicker solid insulation to compensate for the capacitance increase (*Fig. 8a*). The operating companies complained of less pairs per cable sheath.

Next, air bubbles were injected into the plastic to decrease the capacitance, yet maintain a smaller diameter foam insulation around the conductor (*Fig. 8b*). The operating companies complained of the insulation pulling off the conductors during the splicing process.

The manufacturer then added a solid outer skin, forming a foam-skin insulation which was acceptable (*Fig. 8c*).

Each of these techniques compensated for the increase of mutual capacitance, but capacitance tip- or ring-to-ground was increased. Using the standard .083 μF/mile capacitance settings for air-core cable causes errors in readings as the average setting for solid insulation filled cable is .092 μF/mile, and for foam skin is .088 μF/mile.

> **REMEMBER:** Use standard settings only for measuring the *mutual* capacitance in filled cable. When measuring tip- or ring-to-ground, set .092 μF/mile for solid skin, and .088 μF/mile for foam-skin. When finding opens in a section of buried filled cable, calibrate the test set to the known length on a good pair for the exact capacitance.

Special cables. Open meters are calibrated to be used on telephone cables that are 25 pair or larger.

Drops, six pair, 11 pair, 16 pair, LOW-CAP, and other special cables, even though the *mutual* capacitance may be .083 μF/mile, have a different capacitance to ground.

An example of this is the "B" service wire used for buried drop. This wire is identified by the four conductors colored red, green, black, and yellow. The drop is air-core, the shield is aluminum. The mutual capacitance is .083 μF/mile. The capacitance-to-ground is .12 μF/mile.

To find the capacitance of special cables and drops, take a trip to the equipment yard and measure known lengths of this type of plant. Set this known footage into the open meter and adjust the capacitance according to the instructions in the manual. Record this information. When any of these pre-tested conductors are en-

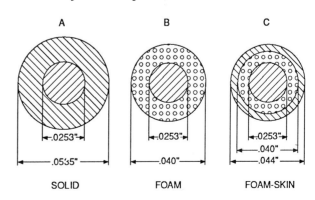

Fig. 8. Comparison of diameters for solid (A), foam (B), and foam-skin (C), insulation for filled cable.

countered in the field, use the capacitance found in the tests.

Water. Finding opens in air-core PIC cable with water present results in false readings. The measurements will be long by a factor of almost three over dry cable measurements.

For example, an open meter is accurately test-calibrated to read 1000 feet of air-core cable. If that same cable is pumped completely full of water and remeasured, the open meter will read 2865 feet of cable, or almost three times longer than the physical length. The extra footage represents a capacitance increase of 1.865 over the actual length of air-core cable without water.

In partially wet cable, this increase exists only in the wet area. To analyze a suspected wet section, measure and record the actual distance of the section. Next, measure the length of an open pair. Subtract and divide the difference by 1.865. If a 700-foot section measures 1100 feet, the 400 foot difference divided by 1.865 equals 214.5 feet of water in the section (*Fig. 9*).

NOTE: *A Time Domain Reflectometer (TDR) can be used to identify and measure the distance to a clean open or a noisy open (0-100 ohms) and will show where water is located in a section.*

Using the TDR

For the purpose of this book, the benefits of a TDR need to be discussed.

A TDR is a companion meter for fault location in copper plant. The set fills gaps in fault analysis and location that cannot be handled by other meters. A TDR is not a stand alone meter.

Advantages of a TDR:

■ The ability of the trace, when properly programmed and interpreted, to indicate where water starts and stops in an air-core cable.

■ Identifies and measures the distance to a load coil (a TDR cannot read through a load coil).

■ Measures the distance to a split cable pair and a split and corrected cable pair (provided no load coils are in series).

■ Identifies the presence of a lateral and measures the distance to the bridge splice.

■ Identifies and measures the distance to clean opens.

■ Identifies and measures the distance to a noisy open (0 to 100 ohms).

■ Identifies and measures the distance to low resistance faults.

■ Identifies and measures the distance to buried splices.

The test set requires a great deal of expertise to operate and interpretations of the readings can vary between sets and operators. When used with a resistance bridge, an open meter, and an earth gradient frame, a TDR is a good tool for a cable repair crew.

Three-terminal open measurements

Some central office mechanized loop test systems and newer field open meters simultaneously measure the individual mutual capacitance between tip and ring, tip and ground and ring and ground (*Fig. 10*).

The sets measure capacitance tip-to-ring,

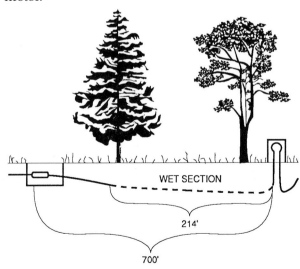

Fig. 9. Water in PIC.

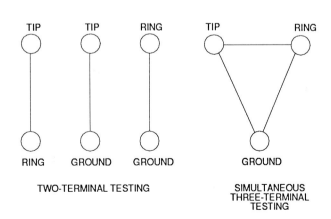

Fig. 10. Two-terminal testing and simultaneous three-terminal testing.

tip-to-ground, and ring-to-ground individually by isolating each portion of mutual capacitance on the circuit. Each such portion is assigned a value, and a comparison is made internally.

This method is accurate when measuring the distance to an open through a cable that is part air-core and part filled. On an unbalanced pair, the sets provide a ratio that allows an accurate distance-to-fault calculation from the far end back. The longer of an unbalanced pair measurement is also acceptable.

There are advantages and disadvantages to each technique whether it is a two-terminal or three-terminal measurement. A well-trained, skilled technician will be able to use either type with excellent results.

Cable capacitance is that electrical property in cable which allows rapid location of opens. It is also the property used by split locating test sets to analyze capacitive loss and indicate split location.

TERMS TO REMEMBER: *(Write the definitions in your own words.)*

Capacitance—...

....

Capacitor—...

....

Mutual capacitance—..

....

Unbalanced pairs—...

....

Time Domain Reflectometer (TDR)—...

....

REVIEW QUESTIONS:

1. What is the capacitance goal of cable manufacturers?
A. .083 μF.
B. .091 μF.
C. .166 μF.

2. Capacitance is affected by:
A. Gauge.
B. Temperature.
C. Neither.

3. Opens in air-core PIC with water present result in measurements that are:
A. Short.
B. Long.

4. A TDR reading passing through a load coil is:
A. High.
B. Low.
C. Blocked.

5. An open meter measures all wire, including any laterals.
A. True.   B. False.

6. The effect of induced AC on an unbalanced pair:
A. Rings, can't answer.   C. Audible noise.
B. No dial tone.   D. Any of the above.

7. If tip and ring measurements differ, use the:
A. Shorter.
B. Average.
C. Longer.

(Answers on page 89)

CHAPTER 8

OUTLINE

Split pairs—the toughest repair problem

Measuring a split
- Different cable types
- Laterals or bridged tap
- Split and corrected
- Toning the split

OBJECTIVES

After completing this chapter, the student will be able to:

1. Identify problems caused by split pairs.

2. Analyze plant for a distance measurement to a split.

3. Know what to do and what not to do when a split is located.

Identifying and Locating Splits

PREVIEW QUESTIONS

As you read, watch for the answers to the following important questions:

1. How do split pairs occur and what is the result?

2. What factors must be considered in identifying a split?

3. How is the split located?

Split pairs are probably the single most difficult fault repair problem in outside plant. They are most time consuming to find, and they are properly left to the desperation of full-count cable complements before they are attacked.

A split occurs when conductors of pairs with different twists are spliced together along a cable route (*Fig. 1*). This is a man-made problem, and usually occurs in a splice—although some occur in cross-boxes and access points.

Because the twist control is destroyed, the capacitive balance of the cable is affected, and the common customer complaint is crosstalk. If the pair is split for a long enough distance, the conversation on the other pair can be heard quite clearly.

The most common solution to the crosstalk complaint is to cut the pair. As the cut pair shows no resistance faults and tests as a good vacant pair, it will be cut back to repeatedly. This will cause further complaints from the same customer, and eventually a Public Service Commission complaint.

NOTE: *When a split is encountered and not repaired, it will test vacant-good. It must be tagged as a universal bad pair to avoid future cutovers.*

Usually the split pair is not repaired until all other vacant pairs have been used, and all other faulted pairs have been repaired and used. Only when the count is full and a pair for a new service order is needed is the split traced.

The first man on site making the installation usually misinterprets the split as one side of the pair open. With no other pairs to cut to, the problem is turned over to the Cable Repair Technician (CRT).

When testing with an open meter, the CRT usually finds a balanced pair, which places any

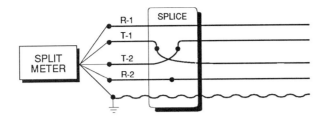

Fig. 1. Split cable pairs.

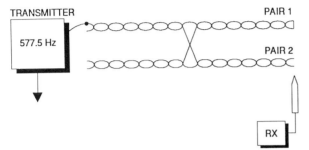

Fig. 2. Identifying a split pair.

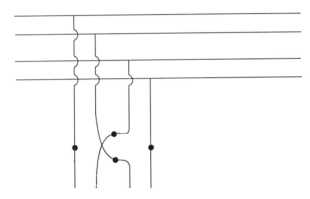

Fig. 3. Split on a lateral.

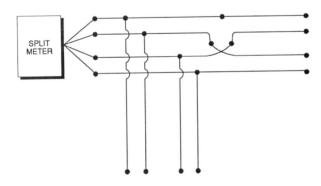

Fig. 4. Subtract length of lateral.

open near the far end. When remeasuring back from the far end, the pair will still read as balanced. The problem is finally diagnosed as a split.

To identify the pair with which the test pair is split, tone (577.5 Hz) is applied. The tone

is identified in the count with an amplifier. The split pair mate shows more tone on it than any other pair in the count (*Fig. 2*).

There are several open/split meters available today. They differ in detail, but all use the capacitance loss beyond the split for location.

Measuring a split

Following is a detailed, step-by-step routine to analyze the plant and obtain distance measurements to the split.

Different cable types. When testing through two types of cable, the split may incorrectly measure to the juncture. The test set sees the change and interprets it as the site of the split because of the drastic capacitance change from pulp 50 and 100 pair groups to PIC 25, 12, and 13 pair sub-units.

When this occurs, separate the cables, determine the direction of the split pairs, and then remeasure the distance to the split. For example, if a pulp cable from the CO is spliced into a PIC cable at a control point, open the pairs and remeasure from the control point.

Laterals or bridged tap. When a single split occurs on a lateral (*Fig. 3*), the pair will measure as if it was split and corrected. Most likely, the far-end-to-split measurement will indicate the distance to the split from the end of the lateral.

If a distance-to-split or far-end-to-split measurement includes a lateral (*Fig. 4*), the length of the lateral must be subtracted to give an accurate distance measurement to the split pairs.

Split and corrected (*Fig. 5*). In the process of splicing or installing service, the field technician will correct the split at the point where it's most convenient, rather than finding and fixing the split pairs. In the old days, the most convenient spot was on the field side of the main frame. If that was risky, the tip splice in the vault was next.

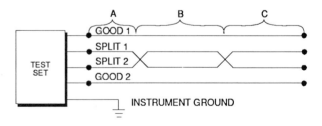

Fig. 5. Split and corrected.

Most field technicians don't realize that as little as five feet of split pairs behind a cross-box will cause crosstalk.

> **REMEMBER:** Never correct a split pair. If no repair can be made, log the pair as a *Universal bad pair.*

The test set will indicate the amount of conductor that is good and the amount of conductor that is split.

As an example, in a 5400 foot section of cable, a splicer split a pair at 100 feet (*Fig. 6*). The problem was identified and corrected at 5100 feet, leaving 300 feet to the end of the cable. With the split meter connected to either end of the cable, the test set would measure a distance to the split of 400 feet (100 good + 300 good). The indicated distance from the far end would be 5000 feet.

The distance to the split measurement is the amount of nonsplit good pair in the test route (here, 400 feet). This measurement, as with any capacitance measurement, includes any laterals.

The far end to the split measurement is all of the split pair (here, 5000 feet), including any laterals.

If the far-end-to-split measurement is longer than the distance-to-split measurement, open the cable at an access near the middle. One split will be in each direction. If the far end to the split is shorter than the distance to the split (*Fig. 7*), find two accesses whose separation matches the far-end-to-split measurement. A split should occur at each one.

With any luck, plant records will show 14 manholes, each separated by the indicated footage. This will allow the technician to practice toning the cable for splits.

NOTE: *A Time Domain Reflectometer can be used to identify and measure the distance to a split cable pair, provided there are no load coils in series with the measurement.*

Toning the split. To identify the exact splice where the split occurs, use a constant tone source, such as the 577.5 Hz breakdown tone, and an exploring coil. With all four conductors of the split pairs strapped together at the far end, apply tone to the two split conductors (*Fig. 8*).

The exploring coil will detect loud and "washed" tone before the splice with the split pairs. The tone will decrease beyond the split

and a "picket fence" type of tone will be heard (where the tone gets loud and soft according to the helical design of the cable). In pulp cable, this will occur every 42 inches (one complete lay of the cable). In PIC this will occur every 36 inches (one complete lay of the cable).

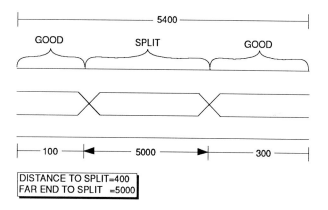

Fig. 6. Measuring split and corrected.

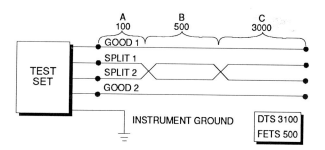

Fig. 7. Connections for a corrected split.

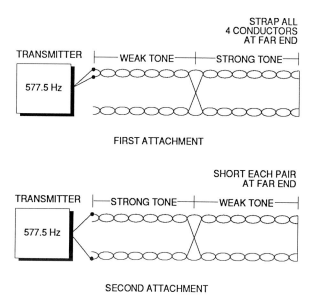

Fig. 8. Coiling a split.

TERMS TO REMEMBER: *(Write the definitions in your own words.)*

Split pair—...
....

Universal bad pair—...
....

REVIEW QUESTIONS:

1. Split cable pairs usually occur in:
A. Access points.   C. A splice.
B. Cross-boxes.   D. All of the above.

2. What is the most common customer complaint resulting from split pairs?
A. Hum on the line.
B. No dial tone.
C. Crosstalk.

3. When applying a tone to identify a split pair, the tone on the split pair will be:
A. Softer.   B. Louder.

4. Split pairs are caused by:
A. Acts of God.
B. Acts of man.
C. Acts of nature.
D. All of the above.

5. A TDR will measure splits accurately as long as:
A. There are at least 12 working pairs in the test cable.
B. There are no loads in the test section.
C. There are no laterals in the section.

(Answers on page 89)

FOR YOUR PERSONAL COMMENTS, OBSERVATIONS AND NOTES:

CHAPTER 9

OUTLINE

Slide wire method

Earth gradient theory

Earth gradient method

Earth gradient test set operation

Finding faults under pavement
- Parallel method
- Extending the frame
- Locating in the parking lot

Faults in Serving Area Concept (SAC) plant

Conductor-to-shield faults
- Metallic contact
- Water in cable

OBJECTIVES

After completing this chapter, the student will be able to:

1. Use an earth gradient test set.

2. Test for conductor-to-shield faults.

Sheath Fault Location

PREVIEW QUESTIONS

As you read, watch for the answers to the following important questions:

1. What tests are used in sheath fault location?

2. How is an earth gradient test set used?

3. What special requirements are there in testing SAC plant?

New problems in section analysis occurred with the use of plastic insulation on cables. With lead sheathing, most cable problems were wet, but swelling of the paper/pulp confined the intrusion to a small area. When polyethylene sheathing was developed, it was glued to an aluminum or steel shield over the paper/pulp conductors. As polyethylene does not block water vapor, the glue itself, not the plastic skin, stopped the intrusion of water. When a failure did occur, the paper/pulp still confined the water to a small area and, of course, the resulting complete cable failure made repairs necessary at once. The fault could be found with breakdown voltage and tone.

As breakdown voltage is useless in PIC cable, finding a problem was more difficult. The presence of water was not immediately obvious and faults occurred slowly over a period of time in the wet section. Bridge and open meters accurately gave distance measurements to the faults. Analysis of their pattern would indicate water, but finding the exact spot of water intrusion (wet splice, sheath damage, etc.) was exasperating and costly. Test holes could become trenches, and fixing a fault in a section could take days. The solution was to cut pairs in the section until there were no spare pairs.

As any sheath damage or splice failure causes an earth ground at the site of the trouble (*Fig. 1*), technicians found that by tracing its gradient, the site could be accurately pinpointed.

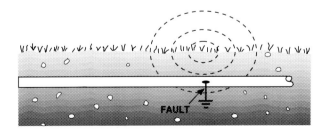

Fig. 1. Resistive fault earth gradient.

The slide wire method

An early procedure to find these grounded conductors was called the "slide wire" method. Two wires were strung on the surface above the bad wire. For accurate measurements, they had to be the same length as the bad conductor and follow its exact route.

A galvanometer was cut into the ring side of the slide wire pair and the pair was connected to each end of the bad wire. A battery was connected to the sheath (negative) and a probe was connected to the positive side with wire long enough to cover the test section (*Fig. 2*).

The tip side of the slide wire pair was probed intermittently until the galvanometer showed no current flow. This was supposed to place the probe within a foot of the trouble below.

This was an arcane and complex process, with many variables which affected its accuracy. The slide pair must exactly match the path and length of the bad wire, which was difficult when locating across traffic lanes and through depth changes. More than one fault in

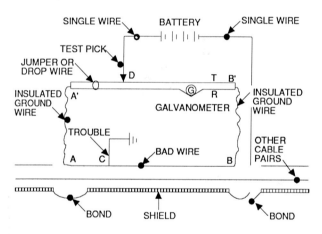

Fig. 2. Schematic diagram, slide wire method.

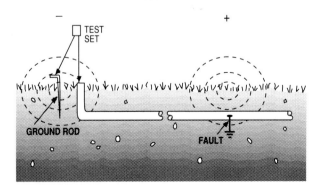

Fig. 3. Gradients at ground and at fault.

the section would show at a midpoint between adjacent faults. As the galvanometer was fixed in place, the procedure required two technicians. Itinerant voltages in the area would mislead the locate. Series resistance in the connectors would affect readings, etc. Confidence was not high when digging.

Because this method was clumsy and time consuming, the need for a test set became obvious. In the 1960s, an earth-gradient set was released which could be operated by one technician. The set worked, but its complexity and varying field conditions rendered it useless to all but the most experienced field technicians.

The need for a simple, accurate test set existed for years. In the 1970s Bell Labs developed two sets, one for buried drops and one for cable in sections.

The sets were buckets. When it rained (the time they were most needed), they were rendered useless. Their high voltage output would overpower the fault if the fault was too close to the set. Also, this voltage had a bit of a bite. Eventually they were repackaged by several vendors, the voltage was made adjustable, and many are in use today. The process was licensed to other vendors and improvements have made the earth-gradient location method a simple and effective task.

Earth-gradient method

The test set applies a voltage to an isolated shield or conductor. A good ground is placed in line with the cable in the direction opposite the fault. When the test set is turned on, two gradients are created (*Fig. 3*). One gradient with positive voltage is at the fault, and the other negative gradient is at the earth ground return.

A receiver, keyed with a higher carrier frequency, uses a two-pronged frame to measure voltage difference between the two probes (*Fig. 4*). A meter on the receiver indicates which probe is detecting the most positive voltage. As the meter would tend to switch back and forth according to the Hz cycle, the negative signal is clipped and only positive voltage present shows on the meter.

The basic operation is quickly learned, even by the most inexperienced technicians.

Earth-gradient test set operation

Following is a step-by-step process to set up and operate an earth-gradient device *after*

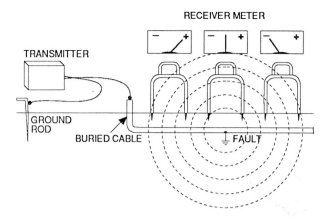

Fig. 4. Meter indication around the fault.

running the self-test to assure the set is in operating condition.

1. Isolate the cable or conductor under test. Remove the ground at both ends of the cable or drop so the shield is floating. If using a conductor, use the one with the heaviest fault, and isolate the conductor and shield at both ends.

2. Use a VOM. Test for shield- or conductor-to-earth fault

NOTE: *Any battery showing on the conductor or shield indicates a possibility of no earth fault in the section under test. In this case, the gradient device would attempt to locate the CO battery. This is not useful unless the technician is lost and needs to find his way home.*

3. Locate the section under test. Use the cable locator portion of the set to locate the path of the drop or section of cable under test. This shows the path to test with the earth frame.

4. Ground the test set. Place a good earth ground opposite the direction of the fault in line with the cable path (*Fig. 5*). Don't use common grounds such as water pipe or electrical conduit. Use an independent ground (a screwdriver or a ground rod will suffice).

5. Set up the transmitter. Use only enough

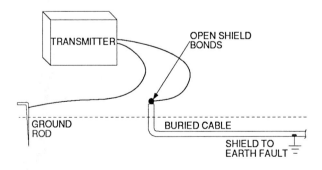

Fig. 5. Transmitter ground placement.

voltage to create a good gradient. Too much can mask the fault. If the test set has no voltage adjustment, add resistance in series to the ground wire. This will decrease the size of other gradients (such as nicks in the cable, or nonfaulted grounded splices tied to the shield) and reduce the chance of their masking the fault.

6. Set up the receiver. Some earth-gradient sets use an audible tone to indicate when to note the needle deflection direction (plus or minus). With these, make a proper volume adjustment at the hookup and leave it, as any later change in volume or gain may confuse the locate. Other gradient receivers do not use tone. With these, the meter deflection alone is observed.

NOTE: *All connections must be in the correct socket and firmly connected. All switches must be in the proper position. All cords and frames must self-test good.*

7. Fault identification. Follow the path of the section or drop under test. When needle reversal occurs, turn the frame at a 90 degree angle and pinpoint the fault (*Fig. 6*). After marking the first fault, follow the complete path of the cable to find any and all other sheath or conductor faults in the section under test.

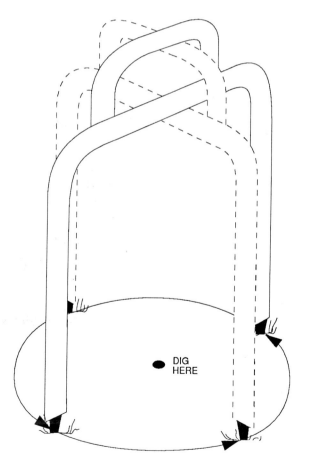

Fig. 6. Frame location, checking location.

Finding faults under pavement

When a cable fault occurs beneath paving, the earth contact frame probes cannot make contact with the gradient directly over the fault. In such a case, there are several methods to accurately pinpoint the fault from a distance.

Parallel method—If a cable runs under pavement the length of the roadway, use the earth frame on the side of the roadway and probe the earth parallel to the cable path until meter reversal occurs. At this point (*Fig. 7*), the frame center is exactly at a right angle to the fault. Use the cable locate portion of the set to locate the cable path and mark where the measurements cross.

To check the accuracy of this measurement, move the frame back along the parallel path a few yards and rotate a few degrees at a time until the meter reverses with only an inch or so of movement. Again, a line marked perpendicular to the frame center will intersect the cable at the fault. (*Fig. 8*). Repeat this procedure a few yards ahead of the original reversal to bracket the measurements for an accurate fault location.

Extending the frame—When a cable passes under a roadway or other relatively narrow stretch of paving and the fault proves underneath, it can be triangulated as above. A more accurate method is to use an extended frame to locate the fault.

Once the cable path is known and marked, place the positive probe in earth at the roadside and hold the negative probe clear. Strip about a foot of an insulated conductor such as cross-connect wire and metallically wrap the exposed negative probe with several turns. Lay out enough wire to equal twice the width of the roadway. Strip the other end and wrap it tightly around a screwdriver. Place the screwdriver in the ground over the cable on the far side of the pavement (*Fig. 9*).

The screwdriver now acts as an extension of the negative probe. Note the meter deflection. If the meter reads negative, the fault is nearest the screwdriver.

NOTE: *When the fault shows to the screwdriver side, reverse the position of the frame and screwdriver to facilitate the locate.*

If the meter reads positive, probe with the positive leg only along the cable path away from the roadway until reversal. At this point, pull the wire tight between the positive probe

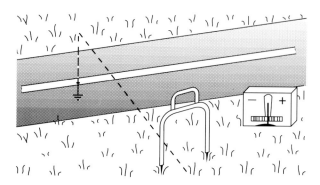

Fig. 7. Fault location perpendicular.

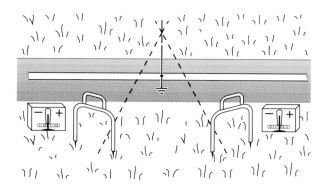

Fig. 8. Fault location triangulated.

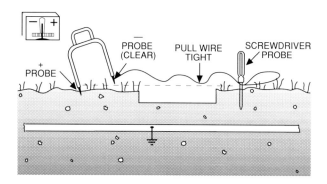

Fig. 9. Frame extension with screwdriver probe.

and the screwdriver. The fault lies exactly half this distance under the pavement. Fold the wire in half and mark the cable path at the fault.

Locating in a parking lot—When a cable is routed under a vast expanse of asphalt, such as a parking lot, mark the path and drive nails, spaced apart the distance between the probes, to penetrate several inches into the earth below. The fault is bracketed using the nails as probe extensions.

This is the way an earth-gradient device was designed to be used and represents an uncomplicated locate. Because field conditions and construction methods have changed since

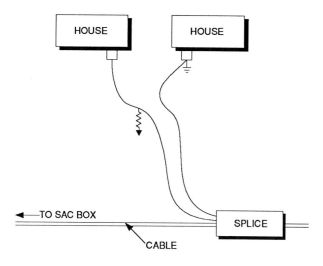

Fig. 10. Drop fault.

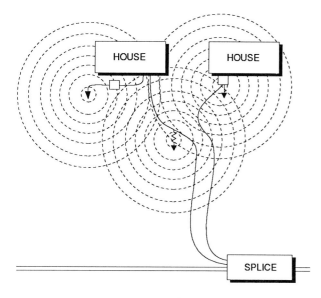

Fig. 11. Gradient mixing.

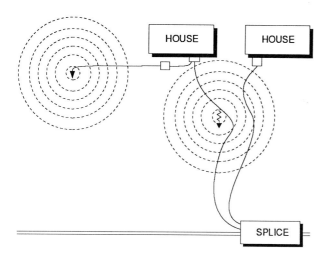

Fig. 12. Area electrically clean.

the devices were invented, special techniques for set-up are now required.

Using an earth-gradient device was easy when operating companies were using ready-access plant. In today's environment, most plant is being locked up. Technicians have to go to great lengths to isolate sections of cable or buried drops. Ground fault locating is time-consuming and in some instances, impossible.

An example of this occurs in the Serving Area Concept (SAC), plant.

Pinpointing a drop fault in SAC plant

Central office resistance bridges and bridges in the field can prove a fault to the drop or the cable, but the exact point cannot be identified. Because of the limited access, a fault in the drop cannot be isolated without extra effort and, in most instances, digging up the encapsulated splice.

An improper hookup can easily mask a suspected drop fault. For example, a faulted drop is in trouble 10 feet from the protector. The drop is short, approximately 60 feet long, and encapsulated directly into the cable. The next door neighbor's drop is in a common trench with the faulted drop. The neighbor's protector is 15 feet away from the fault (Fig. 10).

If the technician connects the sheath fault locator according to instructions in the manual, success is probably nil. First, to properly place the ground (in line with the cable path away from the fault), the technician would have to open the living room window and drive a ground rod through the center of the floor, or go through the entire house, out another window and to ground. When permission for that is difficult to obtain, the next best method appears to be running the ground along the side of the house at an angle.

With this, an absolute electronic mess is created when the test set is turned on. The negative gradient from the ground rod overlaps the positive sheath fault on the drop. The neighbor's protector ground confuses the effort by mixing in the immediate area of the fault (Fig. 11).

Options: place the ground in the neighbor's yard, a good 200 feet from the immediate area of the fault (Fig. 12); next, remove the ground from the neighbor's protector (without interrupting service). This cleans the area electronically and the fault will be easy to locate.

If it is impossible to disconnect the neighbor's ground, place the test set ground as close as possible to the neighbor's ground (*Fig. 13*). As the neighbor's ground appears as positive to the test set, and the set's ground is negative, the two gradients should cancel each other and allow the positive fault to be pinpointed with the frame.

If this doesn't work, the size of each individual gradient must be decreased. To do this, add series resistance (one to five megohms) to the circuit at the ground lead (*Fig. 14*). If possible, cut back on the test set power.

Use the frame and receiver to test the size of the gradient. For example, a 12 to 15 foot negative gradient would indicate that no positive gradient would be larger. The fault should be identifiable, without interference, if no other grounds (grounded splices, power pads, etc.) are within 12 to 15 feet of the trouble.

> **REMEMBER:** Back up all sheath fault locations with a resistance or capacitance bridge measurement to eliminate unnecessary digging.

Conductor-to-shield faults

Conductor-to-shield faults are ground faults which cannot be located with an earth-gradient device. Current flow is through the shield to the proper sheath ground and not to earth at the fault. There are two basic examples of these.

Metallic contact. Kinks can be pulled into new drops. When this happens the continuity between the shield and the sheath is not broken, but the conductor is grounded to the shield (*Fig. 15*). What appears to be a ground is a fault between the conductor and the shield. When testing, the technician interprets grounds from other protectors and SAC boxes as a conductor-to-earth fault. If the technician digs up and isolates the drop and tests with a VOM between the open conductor and earth, no fault is evident on the conductor.

Water in the cable. Water standing in an air-core section of cable creates battery faults between the conductors and the shield. Between 15 and 30 volts DC will appear on the isolated shield when it is tested to earth-ground. This battery can be misinterpreted as an earth-ground when

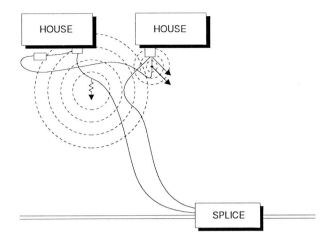

Fig. 13. Concelling effect of neighbor's ground.

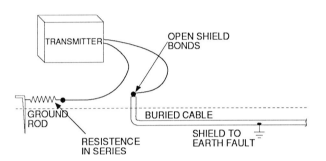

Fig. 14. Transmitter ground placement.

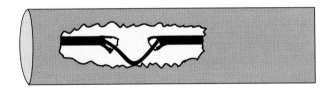

Fig. 15. Conductor to shield fault.

testing resistance. As no battery would appear on the shield if a sheath fault were present, take VOM readings first for voltage. If no voltage is present, a sheath fault is indicated. If voltage is present, there is water in the cable with no sheath damage. In this case, the faults must be located with a resistance/capacitance bridge combination or a TDR.

We have examined the most common types of faults and the electrical principles used in tracking them down. We turn now to the study of section analysis, the thorough study of a trouble, and its complete repair.

TERMS TO REMEMBER: *(Write the definitions in your own words.)*

Conductor-to-shield faults— ....

....

Slide wire method— ....

....

REVIEW QUESTIONS:

1. *In the slide wire method, the two wires strung on the surface must be:*

A. *The same length as the bad conductor.*
B. *Of different lengths.*
C. *It doesn't matter.*

2. *When battery shows while testing the shield for an earth fault, it indicates:*

A. *The location of the fault.*
B. *The possibility of no earth fault in the section.*

3. *If the test set has no voltage adjustment, when necessary, the technician can:*

A. *Take the average of the highest and lowest voltage readings.*
B. *Add resistance in series to the ground wire.*

4. *A sheath fault is present when a VOM reading shows:*

A. *Low voltage.*
B. *High voltage.*
C. *No voltage present and resistance to earth.*

5. *An earth gradient appears at:*

A. *Test set ground.*
B. *All conductor-to-earth faults.*
C. *All shield-to-earth faults.*
D. *All of the above.*

6. *If it is impossible to disconnect an interfering ground:*

A. *Cut to clear.*
B. *Place set ground next to interfering ground.*
C. *Ground test set to station or pedestal ground.*

(Answers on page 89)

CHAPTER *10*

OUTLINE

Step-by-step guide to section analysis

Physical-to-electrical comparison tests

Resistance affecting conditions
- Gauge change
- Slack loop or butt splice
- Load coil
- Temperature setting and temperature change
- Faulty test set

Capacitance affecting conditions
- Water in air-core cable
- Unknown laterals

Section analysis—the movie

A section history

OBJECTIVES

After completing this chapter, the student will be able to:

1. Choose the proper test set for section analysis.

2. Conduct physical-to-electrical comparison tests.

3. Cite the conditions that must be considered in section analysis.

Section Analysis

PREVIEW QUESTIONS

As you read, watch for the answers to the following important questions:

1. What test sets are needed for section analysis?

2. What factors affect resistance measurements?

■■■■■■■■■■■■■■■■■■■■■

Section analysis is the thorough study of a faulted telephone cable. This analysis determines the type of cable, the vicinity of the fault, the probable cause of the failure, and the permanent fix—all before digging up the cable.

Priority one on any trouble report is service restoral. The customer must have dial tone, even if the fix is known to be a temporary one on a deteriorating cable section. Competent section analysis aids in the quick fix, and the economical scheduling of permanent repairs.

Proper section analysis requires a full understanding of cable makeup and manufacture, test set operation, resistance, capacitance and transmission theory, and a good knowledge of telephone company construction methods, both past and present.

A good electronic picture can be established by a well-trained technician within 15 or 20 minutes of opening the accesses and measuring.

The technician should have available the following test sets for complete analysis:

■ A volt ohmmeter
■ A resistance bridge and zero ohm strap
■ A capacitance bridge
■ A cable locator and earth gradient frame
■ Optional: a Time Domain Reflectometer (TDR)

Each set should be self-tested before use.

Following is the step-by-step procedure for section analysis. First, look around. The trouble site may be obvious. If not:

1. Open the section under test. Remove all bonds.

2. Use a VOM to prove the trouble into the section.

3. Test for battery on the shield (indicates water). If no battery is present, test for a shield-to-earth fault.

4. Determine the physical (actual) distance of the section.

5. Measure the length of the section with a resistance bridge.

6. Measure the length of the section with an open meter.

7. Compare all measurements and note any differences.

8. Measure and record the types of faults and distance to the trouble.

9. Locate and mark all sheath faults.

10. Compare resistance measurements with indicated sheath fault locations; dig only where these measurements agree.

11. Fix all troubles, or replace the cable when necessary.

Here is a synopsis of the theory of section analysis:

Physical-to-electrical comparison tests

By comparing the physical length of a test section to its resistive and capacitive electrical length measurements, many characteristics which affect fault measurements can be deduced. If available, a TDR trace can be used to further analyze the section, identify loads and laterals, and help prove the troubles as to type and location.

Once a fault is isolated to a section, determine the actual or physical distance between the two section accesses. Use plant records, sequential markings on the sheath, or locate the exact path and measure the distance with a tape measure or wheel. Do this before any electronic test equipment is used for analysis.

Next, short a pair at one access. Use a resistance bridge to measure the resistive length of the section (distance-to-strap). Be sure the correct gauge and exact conductor temperature are entered. If the distance-to-strap measurement and the physical length are the same, open the pair and measure with the open meter. If all three measurements agree, proceed with the fault locate.

If the open meter measurement and the physical distance agree but differ from the resistive distance, suspect one or more of the following conditions that affect resistance measurements: gauge change; slack loop; load coil.

If the resistance bridge and the physical distance agree, but the open meter reads long, suspect one of the following conditions which

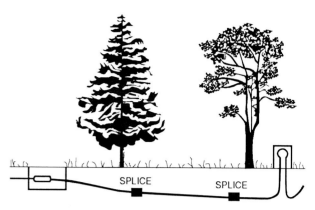

Fig. 1. Gauge change.

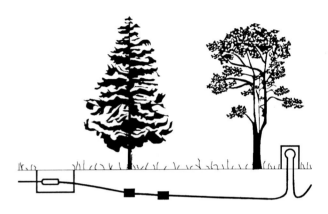

ACTUAL DISTANCE	=325'
OPEN METER MEASUREMENT	=325'
RESISTANCE BRIDGE MEASUREMENT DTS	=362'
DTF	=162'
STF	=200'

Fig. 2. Comparing actual distance to electrical readings.

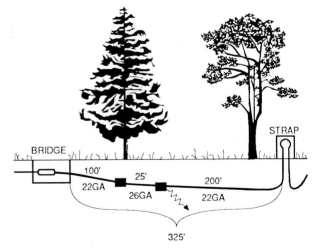

Fig. 3. Fault proved by digging.

affect capacitance measurements: water in air-core cable; unknown laterals.

The presence of one or any combination of these conditions can be confirmed by the physical-to-electrical comparison tests.

Resistance affecting conditions
Gauge Change. If the electrical distance does not agree with the actual distance between the accesses, measure the distance with an open meter. If the open meter measurement agrees with the physical distance, a gauge change is evident, because open meters are unaffected by such a change. If the conductor gauge is the same at both terminals, a gauge change with two splices is indicated (*Fig. 1*). An electrical measurement *shorter* than the actual distance indicates a *larger* gauge cut in, and a measurement *longer* than the actual distance indicates a *smaller* gauge between the splices.

One of the splices is the most probable trouble site. Depending upon which splice failed, the distance to the fault from one terminal, measured through the gauge change, would be wrong. The distance to fault from the other terminal would be correct.

EXAMPLE: We helped a technician analyze a section of 22 gauge cable in Kansas. The physical distance of the section under test was 325 feet (*Fig. 2*). When the far end was strapped, the resistance bridge (set to 22 gauge and the correct conductor temperature) showed 362 feet of cable. The openmeter measurement agreed with the 325 foot physical distance. Suspect: gauge change. Since the resistance bridge measured long, a smaller gauge (24 or 26) was cut in.

Resistance fault measurements showed distance-to-fault as 162 feet, and strap-to-fault as 200 feet. One of these measurements was correct, but which one?

NOTE: *All too often, an inexperienced technician will mark the two measurements and, for some unfathomable reason, dig between them. We refer to this as a "Two-hole case of trouble." First, he digs an I-wonder-if-it's-here "curiosity hole." Next, he opens the cable and remeasures a short distance to the bad splice.*

We were lucky. A shield-to-earth fault locate agreed with the 200 foot strap-to-fault measurement (*Fig. 3*). We dug up the bad splice, confirmed the gauge change as 26 gauge spliced to 22 gauge, and fixed the trouble. The smaller gauge added the extra 37 feet to the distance-to-strap and distance-to-fault resistance measurements.

Had no earth fault existed, we would have dug at the site of either of the two measurements, as one was correct. There is a 50%

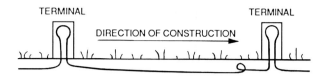

Fig. 4. Slack loop.

chance of being right the first time and a 100% chance of being right the second time.

Had a TDR been available, it would have removed this uncertainty. Properly calibrated to the known section length, the set will show actual distance to the two splices. When these measurements are compared to the bridge measurements, the common measurement is the splice at 200 feet.

Slack loop or butt splice. (*Fig. 4*). If the open meter and the resistance bridge agree on section distance, but the physical distance differs, construction forces may have left a loop of cable on one end of the section or the other. This will mislead a fault location measurement in much the manner of a gauge change. Only one fault measurement will be correct.

When we worked on the line crew, buried cable placement was an educational experience. Engineers have a tendency to survey new placing jobs from a desk. The engineer orders either too little or too much cable for the job.

Often, any excess cable is coiled at the far end and buried. If a slack loop is suspected, consider the logical direction of construction and look for a loop at the end where the crew completed burying the cable.

Sometimes there is too little cable for the section, necessitating an extra splice (often a butt splice) very close to the terminal.

When either a slack loop or butt splice is suspected, use the measurement in the direction of construction to find the fault (*Fig. 5*).

Load coil. An 88 mh load coil adds approximately four ohms to a conductor measurement in a section. Example: a resistance bridge

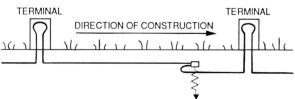

Fig. 5. Butt splice.

measures a 500 foot section of a loaded 24 gauge 400 pair cable as approximately 650 feet (*Fig. 6*).

In the common gauges, the approximate footage added to resistance measurements by loading is:

19 ga.: 4 ohms x 124.24 ft. per each ohm = 497 ft.

22 ga.: 4 ohms x 61.75 ft. per each ohm = 247 ft.

24 ga.: 4 ohms x 38.54 ft. per each ohm = 154 ft.

26 ga.: 4 ohms x 24.00 ft. per each ohm = 96 ft.

An open meter will show about seven feet for the older 632-type load coils and will measure as much as 100 feet for the newer pair-wound 652- and 662-type load coils.

A TDR will show the load coil and indicate the distance to it.

Temperature setting and temperature change. Be sure to set the proper temperature of the conductor in the measured section. Don't guess. If the temperature is set wrong, for each 10 degrees of error on each 100 feet of cable measured, the distance will be wrong by 2.18 feet.

EXAMPLE: Technicians in Arizona were using 70 degrees for underground temperature settings in the summer. We installed a thermometer probe 27 inches in the ground. Buried temperature was 92 degrees at 8:00 A.M. and 96 degrees at 5:00 P.M. This added an error to their measurements of 4.7 feet per hundred in the morning, and 5.6 feet per hundred in the afternoon. When there were no sheath faults present, they were digging a lot of curiosity holes.

Aerial cable temperature can be as much as 40 degrees hotter than ambient temperature. To find the actual conductor temperature, cover a thermometer with black tape and leave it in the sun for at least five minutes.

Faulty test set. A self-test should be run before using any test set for fault measurements in the field. When discrepancies in measurements are noted, re-do the self-test and check the batteries. Errors due to faulty test sets are common.

Capacitance affecting conditions

Water in air-core cable. Water in air-core cable causes an open meter measurement almost three times as long as the true cable section length. A 1000 foot section of air-core cable, full of water, measures 2865 feet capacitively. This increase is 1.865 over the air-core cable with

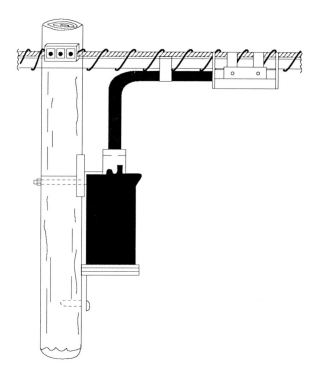

Fig. 6. Load coils add electrical distance.

no water. To estimate the amount of water in the test section, divide the difference of the bridge and open meter measurements by 1.865.

EXAMPLE: The actual section length is 700 feet, confirmed by the resistance bridge. The opened conductor measures 1100 feet with the open meter. The difference, 400 feet divided by 1.865, indicates 214 feet of water in the section (*Fig. 7*). A TDR will show where the water starts and stops in the section.

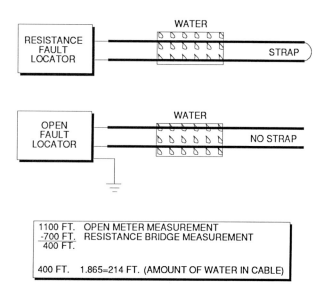

```
1100 FT.   OPEN METER MEASUREMENT
-700 FT.   RESISTANCE BRIDGE MEASUREMENT
 400 FT.

 400 FT.   1.865=214 FT. (AMOUNT OF WATER IN CABLE)
```

Fig. 7. Measuring water in the cable.

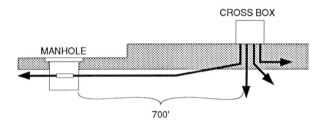

Fig. 8. Faulted section.

The VOM shows the faults as battery on the ring side only, and battery on the shield. The resistance bridge shows faults throughout the length of water in the section.

Intermittent water will result in random resistance fault measurements. Conductors will show faults at irregular distances. If no obvious sheath damage, splices or encapsulations are evident, replace the portion of cable in the section that shows water. A TDR will show the location of each water pocket.

Unknown laterals. As service requirements for an area change it is sometimes necessary to remove a lateral from service. Proper procedure is to open the bridge splice and trim out the lateral. Rather than digging up a buried splice, however, splicers found it more convenient— or were instructed by a supervisor—to remove the U guard at the pole, clear and cap the cable at ground line and replace the U guard. The cable was removed from the records, but there were still working pairs in the leg.

If the end cap eventually goes bad or if the lateral is damaged by construction in the area, resistance faults measure to the three-way splice which no longer shows on the maps.

When comparing the open meter length to the physical length, the presence of a lateral causes a long measurement in the same way as water. In this case, however, the VOM shows grounded shorts and crossed battery. All faults measure to the three-way bridge splice. No battery is indicated on the shield.

An unknown lateral is easily identified on a TDR trace.

Section analysis—the movie
     Wichita, Kansas, 4:30 Friday afternoon. We were in the field with a technician, demonstrating test equipment. The control center reported that 50 customers were out of service. Initial tests indicated a wet or damaged cable. The technician made a brief, profane remark. He had plans to go fishing on Saturday. We hoped to catch a plane home.

The technician isolated the trouble to a section of cable between a cross-box and a manhole. The section was 700 feet of a direct-buried, 24 gauge, air-core, single-sheath cable. We opened the splice in the manhole, proved the trouble into the section, and opened the bonds on each end (*Fig. 8*).

First we tested the shield for battery (water; no sheath damage). Finding no battery present on the shield, we tested for and identified a 5 k ohm sheath fault that was more likely at the same location as the trouble.

Resistance tests: We strung a good pair on the ground between the accesses for our resistance bridge and shorted a pair at the cross-box. The bridge, set to the proper gauge and temperature, measured the distance-to-strap as the physical 700 feet, confirming the actual distance. This showed there was no gauge change, slack loop, butt splice, or load coil in the test section.

Suspect: either water or an unknown lateral.

Capacitance tests: The open meter showed the section length as 1300 feet. Because the section was between a cross-box and a manhole, there was little possibility of an unknown lateral. The excess 600 feet of measurement, when divided by 1.865, indicated at least 322 feet of water in the section.

We tested some 100 conductors with the resistance bridge and graphed the results (*Fig. 9*).

Clusters of faults occurred at 200 feet and at 500 feet from the manhole. The earth frame showed a shield-to-earth fault in both places. We dug at 200 feet and found a splice full of water. The water apparently entered at the damaged spot we found at the 500 foot fault cluster.

We made temporary repairs to restore the service for the weekend. The supervisor made arrangements to replace the cable on Monday. No overtime. (Five bass. One a four-pounder.)

A section history

This section proved to be typical of cable sections nationwide where, through poor analysis or time constraints, technicians work only on the *effect* of a cable trouble. They ignore the cause.

At some time in the past, a stake was driven into this cable, puncturing the shield and damaging several pairs (*Fig. 10*).

Because the fault was in the underground, technicians cut to clear those original trouble pairs. The cause of their troubles—the violated sheath—was not pursued. Over time, water entered the core of the cable and migrated downhill.

As electrolysis set in, more customers complained of noise and service was again restored by cutting to clear. Finally, water entered the splice, and multiple customers were out of service. The section had slowly deteriorated over the 14 months since it was damaged. There had been dispatch after dispatch and cutover after cutover. All due to one small hole in the sheath that should have been found and repaired the first time out.

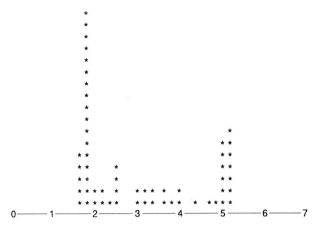

Fig. 9. Graphing resistance measurements.

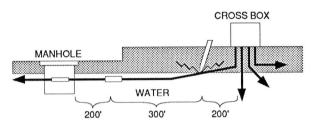

Fig. 10. Cause and effect.

We used section analysis techniques to find the trouble and the source of trouble on a cable which, despite our temporary restoral of service, needed to be replaced at once. If the same simple procedures had been used on the original dispatch to find and repair the cause of those troubles, the cable might well have stayed in service for its intended lifespan.

The cable environment is constantly changing due to growth. New sections of cable on a route may be of another gauge. Pair gain circuits are added daily. It is the responsibility of the field technician to test the transmission characteristics of a circuit at the protector where dial tone is present. We move next to the basics of transmission.

TERMS TO REMEMBER: *(Write the definitions in your own words.)*

Slack loop—..

....

Butt splice—...

....

REVIEW QUESTIONS:

1. *After a fault is isolated to a section:*
A. *Proceed to electrical tests.*
B. *Find actual or physical distance between accesses.*
C. *Dig up the section.*

2. *An electrical measurement shorter than actual distance indicates:*
A. *Smaller gauge cut in.*
B. *Larger gauge cut in.*

3. *When either a slack loop or butt splice is suspected, the fault is located using the:*
A. *Measurement away from the direction of construction.*
B. *Measurement in direction of construction.*

4. *An 88 mh load coil adds about:*
A. *4 ohms.*
B. *2 ohms.*
C. *12 ohms.*

5. *When a resistance measurement reads shorter than the physical distance, and the open meter measurements agree with the physical distance, the condition indicated is:*
A. *Water.*
B. *Load coil.*
C. *Gauge change.*
D. *Slack loop.*

(Answers on page 89)

CHAPTER *11*

OUTLINE

Loop current

Circuit loss

Circuit noise

Transmission pair trouble

Using a circuit termination set

Bonding and grounding

Gain/slope

Testing ground ohms

Designed special service lines

OBJECTIVES

After completing this chapter, the student will be able to:

1. Identify the side of a pair with a series fault.

2. Test for circuit loss and circuit noise.

3. Sectionalize noise problems.

4. Check transmission on special circuits.

Parameters of Quality Service

PREVIEW QUESTIONS

As you read, watch for the answers to the following important questions:

1. What is included in a good loop transmission measurement plan?

2. When are special circuits required?

When installing customer premise equipment, the telco representative recommends various types of lines to connect the customer to the telco's outside plant. Quality standards for the circuits vary according to usage and needs.

The costs of different types of recommended lines also vary, with nondesigned service the cheapest and designed services varying in cost according to the type of transmission quality needed.

Most customers have standard POTS service. At the time of installation, the line must be tested for loop current, loss and circuit noise. These characteristics can be measured and confirmed any time there is trouble in the outside plant that requires a dispatch to the premises.

Any marginal or unacceptable results must be corrected. Problems with a single pair can most likely be fixed on the spot. Bonding or grounding problems, which can affect an entire cable, are referred to a transmission team as necessary.

Business customers use POTS lines for basic service, and in many instances, use the same POTS lines for data transmission.

The customer might get away with this for a while, but there is constant maintenance in the outside plant and the characteristics of the POTS line can change for whatever reason (gauge is changed, etc.). It will still conform to the telco's guaranteed standards, but may not be suitable for the more delicate requirements of high speed data.

On any dispatched trouble, when dial tone is restored to the protector, the same series of transmission tests are made to assure circuit quality.

The telco provides a minimum standard of electrical characteristics for loop transmission. A good Loop Transmission Measurement (LTM) plan measures and records five basic loop transmission parameters:

- Loop current
- Circuit loss
- Circuit noise
- Gain/slope
- Ground ohms

All measurements are taken at the protector, looking back to the CO. A problem with any one or any combination of these parameters will affect analog transmission (which in turn may affect translation to digital circuits). Many of these problems can be rectified by the first man on sight.

We will discuss here what constitutes acceptable transmission, how marginal or unacceptable quality affects service, and the procedures for correcting such problems.

Loop current

Current measurements to dial tone will not measure the actual working loop current because circuit resistance to dial tone is less than to talk battery. Therefore, measurements (in milliamperes or mA) are made to the 1004 Hz milliwatt supply of the CO. This measurement will be the same as the talk battery current supplied to the customer. Portable transmission test sets internally simulate a standard 430 ohms resistance beyond the network interface (*Fig. 1*).

Loop current must be sufficient to provide talk battery and operate supervision and signaling equipment, such as: dial tone request; touch-tone pad operation; ring trip when a call is answered; talk battery to the telephone transmitter.

Insufficient loop current might be responsible for such complaints as: no dial tone; reach wrong numbers; can't be heard; bell rings—can't answer; bell rings after answer.

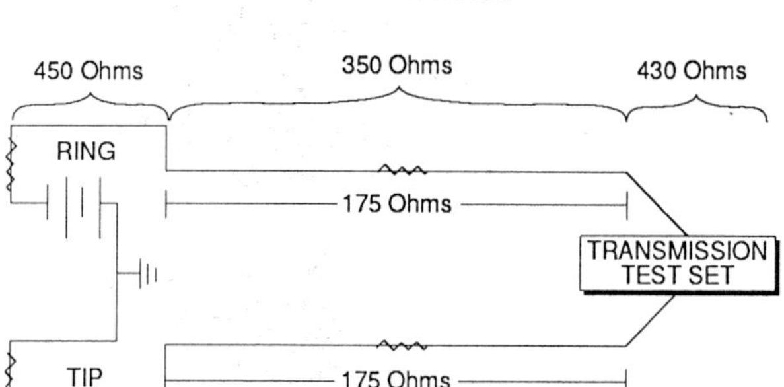

LOOP RESISTANCE

Fig. 1. Transmission.

The *minimum* requirement in most telephone companies is 20 mA into a 430 ohm (subscriber) resistance. This resistance is a combination of 400 ohms for the telephone set, 20 ohms for series devices (MTUs, RIDs, etc.) and 10 ohms for station wire beyond the protector. Any reading below the 20 mA minimum is unacceptable. Any reading from 20 to 23 mA is marginal. Any reading above 23 mA (up to 65 mA) shows acceptable loop current.

When the current falls below a total 23 mA, the telephone transmitter becomes inefficient and problems with low volume are encountered. Marginal current (20 to 23 mA) can usually be corrected by loop treatment which boosts the voltage at the CO. All loops, regardless of length, must measure between 23 mA to 65 mA. Current above 65 mA can cause fading during conversation.

If current falls below 20 mA, additional problems will occur: the CO cannot sense off-hook conditions; dialing problems; ringer problems.

Such extreme low currents can be caused by physical troubles on the line such as a short, ground, cross, battery cross or open. Such faults are identified and located with a resistance or capacitance bridge.

If series resistance faults (wire going open) are suspected, measure a neighbor's loop current. If the neighbor's pair tests acceptable and shows no loop treatment, the trouble is proved a single-pair fault. Use a VOM or resistance bridge to perform a loop and ground test and identify the side of the pair with the series fault. Use the following procedure:

- Short and ground the pair at the CO.

- Measure loop length (ohms or feet) and record the results.

- Measure each conductor to ground.

Ring and tip measurements to ground should be exactly one-half the total loop resistance. If not, subtract the larger ohms measurement from the smaller ohms measurement to determine the amount of series resistance.

EXAMPLE: Loop ohms =1600 ohms. Ring-to-ground =700 ohms and tip-to-ground =900 ohms. There are 200 ohms of series resistance in the tip side. Possibly a bad connector at a splice. More than likely a loose twist.

To determine where the series resistance is,

measure first from the work-out terminal. Next measure from the cross-box. If the pair is balanced at the cross box, the fault is proven between the two accesses. Go halfway between the work-out terminal and the cross-box and repeat the test, moving to the halfway point for each measurement, until the fault is proven to an access or section. (A maximum of seven measurements will find the series fault.) If the fault is in a section, it can be located with a TDR.

If it is not balanced when measuring from the cross-box, the fault is towards the CO and is located in a like manner.

If the circuit is beyond the design parameters of the CO, loop treatment is required.

A REG (Range Extender with Gain) is needed where both loop current and loss are unacceptable. A DLL (Dial Long Line) is needed where current is low and loss is acceptable.

EXAMPLE: a customer would like an Off-Premise Extension (OPE) of a business telephone in another exchange. To boost current in the longer loop, a DLL is added and the loss on the OPE is rectified by an E6-type repeater.

To test for loop treatment, short the pair and measure the voltage to ground with a VOM (*Fig.* 2). The circuit will act as a voltage divider. For instance, A 48-52 volt DC supply from the CO will measure 25 volts DC to ground. A REG will measure approximately 40 volts DC to ground. A DLL will measure 96 volts DC or 0 volts DC, depending on the type.

When loop current is found acceptable, there are several tests that can be made.

Circuit loss

Circuit loss is the signal power lost between the milliwatt supply of the CO and the customer protector. It affects both touch-tone performance and voice volume. Loss tests measure the capacitive characteristics of a circuit in decibels (dB). The amplitude of the signal is attenuated by four primary factors: conductor resistance; insulation loss; capacitance; inductance.

A generated tone of 1004 Hz at 0 dB is placed on the circuit and the attenuation of that frequency from origin to the test set is measured. The maximum length of an unloaded POTS circuit is 18,000 feet. In *Fig. 3*, note that a signal loss to 8 dB is acceptable, from eight to 10 dB is marginal, and above 10 dB is unacceptable.

The capacitance of the entire circuit, including laterals and wire beyond the subscriber,

affects circuit loss and must be included in load engineering.

In areas where there are extremely long loops, it is uneconomical to gauge the cable to bring all subscribers to CO resistance design limits. In such cases, proper loop treatment to increase current will bring the circuit into conformance.

When a marginal or unacceptable loss is encountered, the entire circuit is affected, including bridged tap and wire beyond the customer.

If such a condition is suspected, test the neighbor's pair. If loss is acceptable there, test both pairs for loop treatment. If the acceptable pair is treated, order treatment for the trouble pair.

If no loop treatment shows on either pair, suspect excessive bridge tap or excessive wire beyond the customer on the trouble pair. Remove any such wire beyond the workout terminal and retest. If still not acceptable, look for extended bridge tap. Use the cable map to iden-

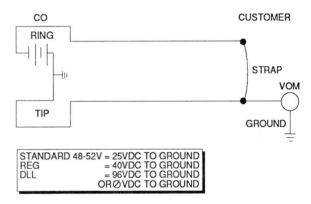

Fig. 2. Testing for loop treatment.

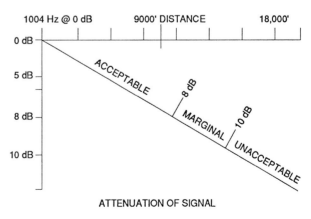

Fig. 3. Nonloaded loop loss.

tify any multiples and, if possible, disconnect the pair at the multiple splice.

> **REMEMBER:** All loop transmission measurements (acceptable or not) must be entered in the telco's centralized data base for total transmission analysis by the transmission engineer.

Only when loss is found acceptable can the next parameter be tested.

Circuit noise

Audible noise on a circuit can be caused by such physical factors as pair trouble, an open lateral, water or bad splicing. If all physical factors test OK, noise is most probably caused by induced AC current from adjacent power lines (*Fig. 4*). This is often called "line borne" noise and is caused by an imbalance in the capacitance between the ring and tip of the pair.

If ring and tip capacitance is balanced, no current flows to ground as the AC current flow is equal and opposite, thus cancelling. Any unbalance, either resistive or capacitive, will cause current—equal to the difference between the two conductors—to flow to ground and "noise-up" the circuit.

Noise Metallic (NM), Power Influence (PI), and circuit balance are interrelated, each being dependent upon the others to form acceptable balance. A formula determines the balance:

$$PI - NM = Balance$$

Noise metallic should be 20 dBrnC or less. Power influence should be 80 dBrnC or less. Therefore, circuit balance must be 60 dB or greater to be acceptable.

NOTE: *Balance calculations are only valid when power influence is 70 dBrnC or greater.*

When noise metallic is marginal or unacceptable and power influence is acceptable, suspect a pair problem. The pair is unbalanced either resistively (going open) or capacitively (one side open on a lateral, beyond the workout terminal, or crossed with a nonworking pair). In most instances, a capacitive unbalance can be identified and isolated with an open meter. If not, use the location techniques for finding a resistive unbalance.

When both noise metallic and power influence are marginal or unacceptable, suspect a grounding, bonding, or associated power company problem (open capacitor bank, bad transformer, open neutral, etc.).

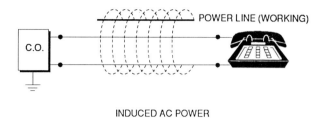

Fig. 4. Line borne noise.

Locating noise problems has always been a confusing process for the technician. It seems that by doing several seemingly unrelated tasks, the noise finally goes away without the basic cause being found.

For example, a technician may remake every ground and bond on an aerial distribution cable. When this task is finished, the noise metallic is acceptable, but no single bond or ground has been identified as the culprit.

Transmission pair trouble—the movie

We accompanied a technician to a nudist colony in the mountains after complaints of a hum on their line. This was very much a back-to-nature establishment, and the nearest power lines were five miles from the site on the CO side of the cross-box. They had telephone service for emergencies only (severe sunburn, etc.).

At our urgent suggestion, the technician decided to go to the work-out terminal instead of the cross-box first. After changing into less formal attire to conform with local custom, the technician dialed up a quiet line circuit and measured the pair. After 50 minutes of testing (the protector was next to the volleyball court), noise metallic measured 25 dBrnC and power influence showed 70 dBrnC, giving a balance of 45dB. With the subscriber equipment removed, the line measured the same (*Fig. 5*).

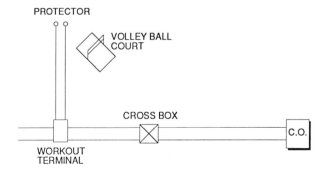

Fig. 5. Difficult transmission problem.

Our next step was to test from the cross-box. After opening the pair, the technician tested back to the office. Both noise metallic and power influence were acceptable. By opening the pair, the AC was removed from the distribution section and the circuit to the CO was balanced. As soon as the section was reconnected, the AC passed into the distribution plant causing the noise to go up.

After several trips into the colony (volleyball: natives 15, technicians 0), we reluctantly pulled out the bridging transformer and quickly proved that the unbalance was on the distribution side and the induced AC was coming from the feeder plant. We isolated the unbalance to an open conductor 100 feet beyond the work-out terminal and the line was cleared. On the trip back down the mountain, the technician complained that her tool belt had badly chafed her skin.

Using a circuit termination set

A circuit termination set (bridging transformer) blocks DC flow while allowing AC to pass uninterrupted. Thus a section can be opened for testing and the AC from the rest of the circuit can pass into the isolated section. Noise problems can be sectionalized in this manner.

When testing a noise trouble, test first at the protector. Remove the station equipment and retest. A drop of three dB or more noise metallic indicates the trouble on the disconnected side.

If noise metallic remains the same, test at the work-out terminal. A three dB or greater drop here indicates a bad drop or the problem is beyond the work-out terminal.

Open the pair beyond the work-out terminal and retest. If still no change, test at the cross-box using a circuit termination set to maintain AC flow. Sectionalize the fault to an access or a section. If the fault is in a section, use a TDR for location.

Bonding and grounding

Marginal or unacceptable grounding, bonding, or associated power company problems are isolated in much the same manner. At least three circuit termination sets are needed. Place one at the protector, one at the cross-box, and one in the CO. This isolates each section from DC influence, yet allows the AC to flow through the entire section.

The fault can be sectionalized with a transmission test set and a circuit termination set. High power influence areas are noted and bonding, grounding, and associated power company problems can be identified and repaired.

Gain/slope

Slope is the graphed difference in circuit loss between the high and low frequencies of a circuit. Lower frequencies (around 1000 Hz) control loudness. Higher frequencies control circuit quality—how well it sounds. If slope is incorrect, the customer can't call out at times, reaches wrong numbers or can't hear at times.

Simply talking over the circuit will not indicate acceptable slope. Higher frequencies are more affected by capacitance than lower frequencies so women and children might find the circuit unusable, while a lower frequency male voice finds no problems.

Loading is a design technique for adding lumped inductance at regular intervals to counter the effect of distributed capacitance. Load coils are used at an exact spacing formula in any circuits longer than 18,000 feet. The first load coil is normally placed at 3000 feet from the CO, and at each interval of 6000 feet thereafter. When a call is placed, the called party's circuit also has a load 3000 feet from the CO, maintaining the 6000 foot spacing. Each subscriber must be less than 10,000 feet from the last load.

If the customer complaint is "can't call out" or "get wrong numbers at times," a possible loading problem exists.

If a load is missing, not placed at the right interval, or if a bridge tap is loaded, circuit loss is increased. Touch-tone dialing problems may be encountered.

Circuit loss is measured at 1004 and 2800 Hz to determine slope. An increase over 7.5 dB between the readings shows unacceptable loss and indicates a loading error. Loading reduces the loss across the voice band (300 Hz to 3000 Hz) by flattening the frequency response curve (*Fig. 6*).

Bridged tap or excessive pair beyond the customer can cause the 7.5 dB level to be exceeded. Loaded bridged tap will reflect the signal back down the pair and cause excessive loss at two frequencies. If these are in the touch-tone band, dialing problems will occur.

These frequencies are tested on a five or

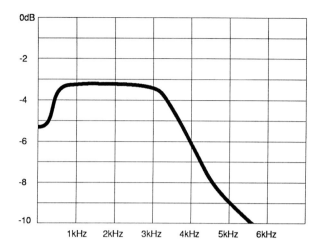

Fig. 6. Slope properly loaded.

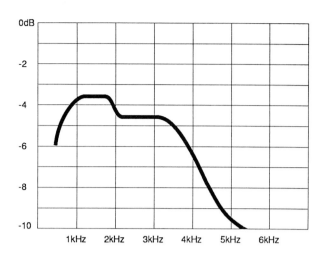

Fig. 7. Missing, reversed, or double load.

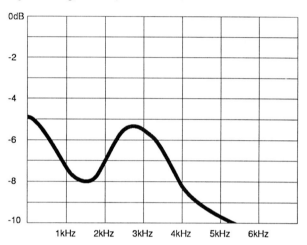

Fig. 8. Loaded or extended bridged tap.

SUBSCRIBER GROUND TEST				
A RESISTANCE LOOP	B APPROX. LENGTH (FEET)	C LOOP (L1) MA	D GROUND (LG) MA	E RATIO LG TO L1
1320	40,700	25	44	1.76
1000	30,800	30	53	1.7
771	23,800	35	59	1.68
600	18,500	40	66	1.65
466	14,400	45	73	1.62
360	11,100	50	79	1.58
272	8,350	55	85	1.54
200	6,100	60	91	1.51
138	4,250	65	97	1.49
86	2,650	70	102	1.45
40	1,200	75	107	1.42

Fig. 9. These limits were established by deriving a mathematical ratio of ground current (LG) to loop current (L1) utilizing the common central office battery.

nine step tone generator. With the results graphed, the response should be flat from 400 Hz to 2800 Hz, then taper off rapidly. A response shown in *Fig. 7* indicates missing, reversed or double loading. A graph as in *Fig. 8* shows loaded or extended bridged tap.

Loading problems are turned over to the transmission engineer.

Testing ground ohms

Use a transmission test set to measure the resistance between the CO ground and the customer ground. The station ground should test 25 ohms or less to assure the carbons at the protector will fire when power or lightning is present on the circuit and protect the customer and station equipment from damage.

Loop current is measured using the 1004 Hz tone. Short the pair (use a screw driver or strap) to hold the circuit up, and move the tip lead of the transmission test set to ground. Remove the short and measure the current from the ring side to ground. The chart in *Fig. 9* indicates acceptable station ground.

If the ground at the protector measures more than 25 ohms or below the minimum levels for that loop, the technician must remake or replace the station ground.

These are the basic parameters for POTS lines. Lines which meet these requirements will support all frequencies up 16 kHz, up to 1200 baud modem signals, fax transmission, card readers, etc. Special service lines are required for safe data transmission at higher rates. Data access lines will transmit a 2400 baud rate, point-to-point lines can work at 9600 baud rate,

and telemetering lines are custom designed for the type of equipment they service.

Designed special service lines

The basic parameters for design services are the same as for any subscriber phone. Special circuits are ordered independently and are specially configured for customer requirements. In addition to measurements for POTS circuits as described above, additional tests assure clean data transmission. The following measurements are required on all designated data circuits.

Peak-to-Average Ratio (P/AR) is a test that simultaneously evaluates the gain and phase distortion characteristics of a channel. It indicates how well the signal comes through the circuit. The P/AR procedure is simple, fast and a single number measure of the overall quality of a data circuit.

C-notched noise is a test that measures the amount of noise on a channel with a signal present. Using filtering, this test detects singing, listener echo and quantizing noise, which may affect data transmission.

Impulse noise is tested over a given period of time (usually 15 minutes) to detect impulse noise peaks which exceed a given threshold.

Phase jitter is normally tested only on the carrier portions of a designed circuit, but end-to-end facility measurements may determine a data trouble to be facility- or equipment-oriented.

Envelope delay distortion indicates the difference in time required for each frequency in the band to travel from one end of the circuit to the other. This impairment is calculable and is controlled by design. Major contributors are analog multiplex equipment and long lengths of loaded cable.

Intermodulation distortion measures the effect of nonlinear elements on data signals. This impairment must be measured on all high performance ("D" conditioned) data circuits.

Return loss is the difference in dB between the power of a transmitted signal on a circuit and its reflected power (echo) which causes impedance mismatch. The most predominant offender is the hybrid at four-wire to two-wire conversions.

The future

Well maintained copper plant is proven good stuff. It is placed with all good intentions and expected to have a life of some 40 years. There is no reason that it cannot reach that expectation—if it is maintained intelligently and correctly. Copper plant eventually will be replaced with fiber optic, cellular and satellite technologies.

This plant is already economically competitive for new construction in the underground plant where installations over 4000 feet are going to fiber. It's cheaper now and cheaper later. High-count copper cable installations eventually require additional central offices, need constant air pressure maintenance, and are susceptible to lightning and induced power. Fiber has none of these drawbacks.

Large office complexes are the next area for fiber. Their needs are expanding to the point where service with fiber optic networks will price copper out of the business environment.

Fiber in the distribution loop depends on several factors. As new uses for residential telephone service develop, such as high resolution television (which requires fiber), the new revenues generated will justify fiber optic cable directly to the home.

Until that time, copper distribution plant must be maintained by enlightened management and highly-trained technicians.

Management must be schooled in the guidelines for economical copper loop installation and maintenance. A tremendous responsibility is placed on the engineering and construction forces in the design of new plant. A varied and hostile plant environment requires the proper cable to be ordered for specific areas of use. This may increase initial cost, but in the long run, high quality cables pay for themselves.

In the long run, high quality anything pays for itself, and the quality of the technical craft is the major cost in outside plant. When the craftsman is knowledgable, confident, trained and ready, the plant will function as it was designed. We hope this book will assist in that goal.

TERMS TO REMEMBER: *(Write the definitions in your own words.)*

Circuit loss—...

....

Line borne noise—...

....

Slope—...

....

Loading—...

....

Return loss—...

....

REVIEW QUESTIONS:

1. The minimum acceptable loop current is:
A. 14 mA.   B. 23 mA.   C. 40 mA.

2. Maximum acceptable station ground resistance is:
A. 45 ohms.   B. 10 ohms.   C. 25 ohms.

3. When NM is unacceptable and PI is acceptable, suspect:
A. Series resistance on the pair.
B. One side open on a lateral.
C. One side open beyond the subscriber.
D. Cross with a nonworking pair.
E. All of the above.

4. Load spacing for an 88 mh load coil is:
A. 1600 ft.
B. 4500 ft.
C. 6000 ft.

5. The amplitude of a telephone signal is attenuated by:
A. Gauge changes.
B. Load coils.
C. Bonding.
D. Capacitance.
E. All of the above.

(Answers on page 89)

ANSWERS TO REVIEW QUESTIONS, WITH COMMENTS

Chapter 1 (Review questions on page 13)

1. A. *(A good conductor of the same gauge is acceptable in a one pair hookup with modern bridges.)*

2. A. *(Breakdown voltage should never be used in PIC.)*

3. B. *(New versions of the Wheatstone Bridge have been developed for troubleshooting in PIC.)*

Chapter 2 (Review questions on page 19)

1. B. *(Because of the helical cable design.)*

2. A. *(Staggered twist improves cable transmission characteristics.)*

3. C. *(Grounded pairs mimic the capacitive effects of working pairs.)*

4. B. *(This is useful when toning splits.)*

5. A. *(Unless calibrated to a known length, open measurements can vary.)*

6. A. *(Cross talk is most often caused by split pairs.)*

Chapter 3 (Review questions on page 27)

1. C.

2. A.

3. B. *(A new technology uses ionization to locate in wet pulp without the dangerous current of the breakdown set.)*

4. B.

5. A. *(Such a section will have to be replaced.)*

6. B.

7. B.

8. C. *(This will eventually corrode the shield open.)*

Chapter 4 (Review questions on page 35)

1. A.

2. B. *(Tinsel, wire, pliers, etc., add series resistance to measurement.)*

3. B. *(Because of the lower resistance of the larger gauge.)*

4. E. *(This equals sheath footage to the fault.)*

5. D. *(The good pair can be any gauge, any temperature, longer or shorter, and not affect the measurement.)*

Chapter 5 (Has no review questions)

Chapter 6 (Review questions on page 48)

1. A.

2. B.

3. B. *(Because the fault beyond measures to the three-way.)*

4. B. *(Current takes the most direct route between the set and strap.)*

5. B.

Chapter 7 (Review questions on page 57)

1. A. *(Capacitance can vary from .076 to .090 μF/mile.)*

2. C.

3. B. *(As capacitance is increased by the presence of water, footage measurements also increase.)*

4. C. *(A load allows frequencies up to 15 kHz to pass; a load coil is operated above 15 kHz.)*

5. A. *(A capacitance measurement includes all conductors in its path.)*

6. C.

7. A.

Chapter 8 (Review questions on page 62)

1. D. *(Splits are always man-made.)*

2. C. *(Crosstalk can occur with as little as five feet of split wire.)*

3. B.

4. B.

5. B. *(Because of this, a TDR will accurately measure the distance to a load.)*

Chapter 9 (Review questions on page 71)

1. A. *(This method is ineffective and is no longer in use.)*

2. B. *(The shield must be isolated for this test.)*

3. B. *(This will decrease the size of the earth-gradient.)*

4. C. *(This fault can be pinpointed with an earth frame.)*

5. D.

6. B. *(The gradients of the two will cancel.)*

Chapter 10 (Review questions on page 79)

1. B.

2. B. *(This can be confirmed with an open meter.)*

3. B.

4. A. *(It adds 8 ohms to a loop measurement.)*

5. C. *(There is at least one splice in the section. If the same gauge appears at both accesses, there are two splices.)*

Chapter 11 (Review questions on page 88)

1. B.

2. C. *(See Outside Plant magazine, July 1989, pg. 12, for detailed discussion.)*

3. E.

4. C.

5. D.

Index

Continue your career development training with other abc aids

The abc TeleTraining manuals have been providing Understandable Technology™ to people in the telecommunications industry since 1942. In addition to the list of training manuals in various series, shown on the copyright page of this book, some two dozen abc training courses have been developed in recent years to meet the need for specialized, concentrated training know-how in specific technical areas.

Instructor-led courses are available as on-premise training programs at your company's site, or in an exclusive abc Flexi-Course format (to meet your personal time and location availability), and on scheduled dates in various public locations around the country. Subjects covered in some of the training courses offered by abc are included in the lists below.

Audio cassette albums include the same complete workbook used in the workshop programs. They include 6-12 tapes transcribed and edited to provide essential training information from the live courses.

Videotapes vary individually from 15 to 59 minutes or more in length, depending on the subject. The workbooks included with these were developed to match the video modules and are instructionally-designed to measure abilities of students to meet the performance objectives of each module.

New courses are being added to the Workshop Course roster continually. For further information, call abc TeleTraining. Customer service desk: 800-abc-4123, or 708-879-9000.

Audio cassette albums:

Fundamentals of Telecommunications
Digital Transmission and Switching
Cellular Mobile Radio Telephone Systems
Practical Signaling Principles
Telecommunications Transmission Systems
Transmission & Signaling Design for Special Services
1A2 Key Telephone Installation (AudioManual)

Principles of Traffic and Network Design
Telephone Technology and Practice
Traffic Basics
Fiber Optic Communications
Microwave Radio System Engineering
Basic Telephone Network Perspectives
Satellite Communications & VSATS, 1-8

Videotapes:

Anatomy of Telecommunications (18 modules)
Digital Transmission and Switching (15 modules)
Telecommunications Transmission Systems (9 modules)

Computer-Based Training (CBT) disks:

Telephone Technologies
Business Systems & Services
Understanding the Public Network
Computers and Data Communications
The TeleCompendium (All 4 courses above)
LAN—Local Area Networking
X.25 in Modern Data Communications

T1 Transmission Basics
ISDN
SNA—System Network Architecture
Data Network Troubleshooting
PWAC Engineering Economics
Instant Traffic Tables
Traffic Engineering Analysis

Home Study Programs:

Many of the abc manuals are available as part of abc's expanding home study testing program, which includes a final exam and certification of completion. To date, the following are available:

The Fine art of Fault Locating
1A2 Key Telephone Installation
Transmission Theory
Microwave Facilities & Regulations
Teletraffic Concepts in Corporation Communications

Practical Grounding Theory
Station Installation & Maintenance
Telephone Theory, Principles & Practice
Introduction to Telecommunications
Principles of Traffic & Network Design